Innovations in Ultra-Wideband Technologies

Edited by Albert Sabban

Published in London, United Kingdom

IntechOpen

Supporting open minds since 2005

Innovations in Ultra-Wideband Technologies
http://dx.doi.org/10.5772/intechopen.92524
Edited by Albert Sabban

Contributors
Gopikrishna Madanan, Deepti Das Krishna, Abdulhameed Habeeb Alghanimi, Chandrasekhar Rao Jetti, Venkateswararao Nandanavanam, Jose Pino Ortega, Markel Rico González, Usha Gautam, Tarun Kumar Rawat, Albert Sabban

First published in London, United Kingdom, 2021 by IntechOpen
IntechOpen is the global imprint of INTECHOPEN LIMITED, registered in England and Wales, registration number: 11086078, 5 Princes Gate Court, London, SW7 2QJ, United Kingdom
Printed in Croatia

British Library Cataloguing-in-Publication Data
A catalogue record for this book is available from the British Library

Additional hard and PDF copies can be obtained from orders@intechopen.com

Innovations in Ultra-Wideband Technologies
Edited by Albert Sabban
p. cm.
Print ISBN 978-1-83968-336-7
Online ISBN 978-1-83968-337-4
eBook (PDF) ISBN 978-1-83968-338-1

For EU product safety concerns:
IN TECH d.o.o., Prolaz Marije Krucifikse Kozulić 3, 51000 Rijeka, Croatia, info@intechopen.com or visit our website at intechopen.com.

Meet the editor

Dr. Albert Sabban holds a Ph.D. in Electrical Engineering from the Faculty of Electrical and Computer Engineering at the University of Colorado at Boulder (1991), and an MBA from the Faculty of Management, Haifa University, Israel (2005). He also holds a BSc and MSc, both Magna Cum Laude, from the Electrical and Computer Engineering Faculty at Tel Aviv University. He is a senior lecturer and researcher in electrical and computer engineering at several colleges. From 1976 to 2008 he worked at RAFAEL as a senior researcher, group leader, and project leader. From 2007 to 2021 he worked as an RF specialist project leader at Hitech companies and a senior lecturer and researcher at various colleges. He has published more than 100 research papers and holds patents in the United States. He has written and edited nine books on compact wearable systems and green technologies. He has also written books on electromagnetics and wideband microwave technologies and six chapters on wearable printed systems and green technologies.

Contents

Preface

Wireless communication systems demand higher data rates, enhanced quality of service, and more channel capacity. These features can be achieved by using Ultra Wideband technologies (UWB), which have attracted considerable attention in the last decade. The inherent features of UWB technologies are a high data rate, more channel capacity, extremely less power consumption, and low cost. Compact wide-band RF modules and antennas are crucial in developing UWB direction-finding systems, radars, seekers, and communication systems. UWB systems suffer from losses due to multipath fading and frequency interference. Input Multiple Output (MIMO) technology can solve multipath interference and losses to improve the dynamic range and channel quality of the communication system.

This book discusses innovation in UWB technologies and systems. Each chapter presents important information to enable engineers, students, and scientists from all areas to follow and understand the topics presented in the book. The book is divided into four sections that present the main topics of green technologies.

1. "Introduction to Ultra-Wideband Technologies"
2. "Ultra-Wideband Systems and RF Modules"
3. "Applications of Ultra-Wideband Technologies"
4. "Ultra-Wideband Antennas and Devices"

Chapters in these sections examine such topics as UWB MIMO antennas and UWB RF modules and systems. Minimization of the size, cost, and weight of UWB RF modules and antennas is achieved by employing MMIC, MIC, and MEMS technologies. However, the integration of MMIC, MIC, and MEMS components and modules raises technical challenges in efficiency, accuracy, and tight tolerances. Accurate design of microwave modules and antennas is a must in the development of UWB systems, as it is impossible to tune microwave devices in the production line. The book also includes chapters on UWB technologies for medical and sports applications and time-domain performance evaluation of UWB antennas.

I want to thank all the chapter authors for their excellent contributions.

Albert Sabban
ORT Braude College,
Israel

Section 1

Introduction to Ultra-Wideband Technologies

Chapter 1

Introductory Chapter: Ultra-Wideband Technologies

Albert Sabban

1. Introduction

Minimization of the size, cost, and weight of the UWB RF modules and antennas is achieved by employing MMIC, MIC and MEMS technologies. However, integration of MIC, MMIC and MEMS components and modules raise technical challenges such as efficiency, accuracy, and tight tolerances. Design consideration and tolerances that can be ignored at low narrow band frequencies cannot be neglected in the design of UWB integrated RF modules. Advanced RF design software, such as ADS, CST, HFSS and AWR, should be used to achieve accurate design of UWB microwave communication devices in mm-wave frequencies. Accurate design of microwave modules and antennas is a must in development of UWB systems. It is an impossible mission to tune microwave devices in the production line.

Design of wideband UWB RF modules, filters and antennas are presented in [1–12]. Wideband RF technologies such as MIC, MIMIC and MEMS are presented in [1–7]. Wide band RF modules are crucial in the development of Direction finding, DF, systems. A fully integrated 10–40 GHz superheterodyne receiver frontend using a 40–46 GHz IF is presented in [8].

Wideband RF technologies are used to develop wideband RF modules such as frontends, active antennas and receiving and transmitting channels as presented in [1–15].

Communication and radar industry in mm wave are currently in continuous growth. The demand for wide bandwidth makes the Ka-band attractive for future commercial communication and radar industry. ADS, HFSS, AWR, and CST are system and electromagnetic software used to develop wideband RF systems, modules, and antennas, as presented in [16–19].

2. MIC and MMIC microwave and MM wave technologies

Compact low cost UWB systems may be developed and manufactured only by using miniature MMIC and MIC components.

2.1 MIC-microwave integrated circuits devices

Communication RF devices and systems consist usually of connectorized modules (such us Mixers, Amplifiers, Filters, and circulators) connected by cables. Connectorized devices are not compact and have big volume. They suffer from high losses and high weight. Volume, weight, and losses may be reduced by using Microwave Integrated Circuits, MIC technology. **Figure 1** presents a MIC Transceiver. MIC devices, standard MIC and miniature HMIC are well known types of MIC devices. Hybrid Microwave Integrated Circuit is named as HMIC device. In MIC design active and passive components are soldered or bonded to the dielectric substrate.

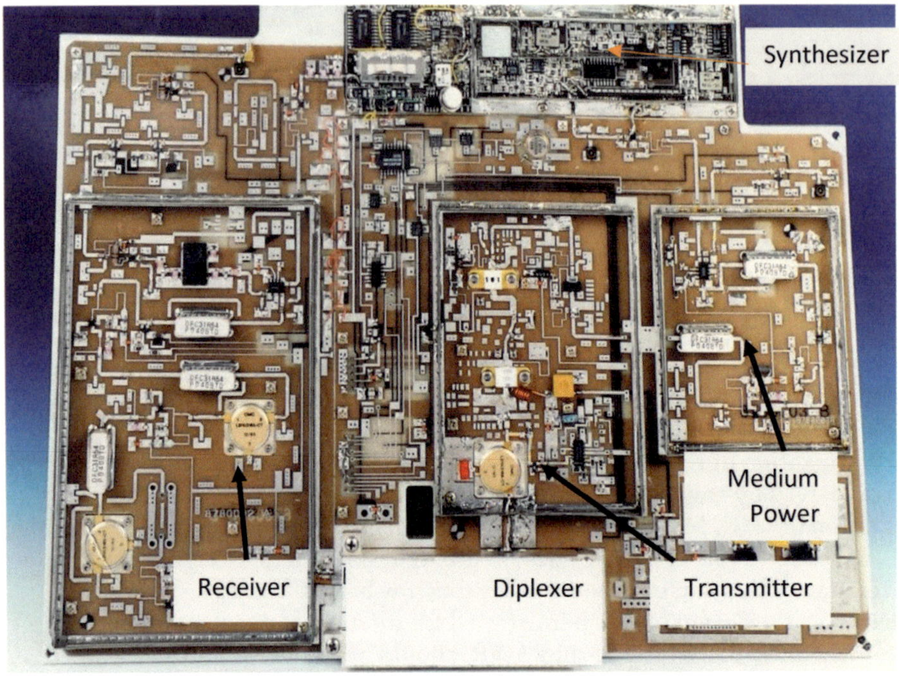

Figure 1.
MIC transceiver prototype for INMARSAT-M ground terminal.

The capacitors, resistors and other passive elements are produced by using thin or thick film technology. A single level metallization for conductors and transmission lines is used in Standard MIC technology. Multilevel process in which passive elements such as inductors, resistors, capacitors, and passive attenuators are batch deposited on the substrate in Miniature HMIC technology. Active components such as mixers, amplifiers and diodes are soldered or bonded on the substrate.

2.2 MMIC- monolithic microwave integrated circuits

MMIC are circuits in which passive and active elements are generated on the same dielectric substrate, as presented in **Figure 2**, by using a deposition scheme as epitaxy, ion implantation, sputtering, evaporation, and diffusion. The layout of the MMIC chip in **Figure 2** consists passive and active elements such as resistors, capacitors, inductors and FET, Field Effect Transistor.

2.2.1 MMIC design features

MMIC components cannot be tuned. Accurate design is crucial in the design of MMIC circuits. Accurate design may be achieved by using 3D electromagnetic software such as ADS and HFSS.

Materials used in the production of MMIC chips are SiGe, Silicon, GaAs, GaN, and InP. MMIC design is sensitive to large statistic scattering of the components, electrical parameters. Production of MMIC wafers in FAB are expensive, around $200,000 per run. Designer goal is to meet with customer specifications in the first design iteration. Compact MMIC components.

yields lower cost of the MMIC chips. **Figure 3** presents MMIC design process.

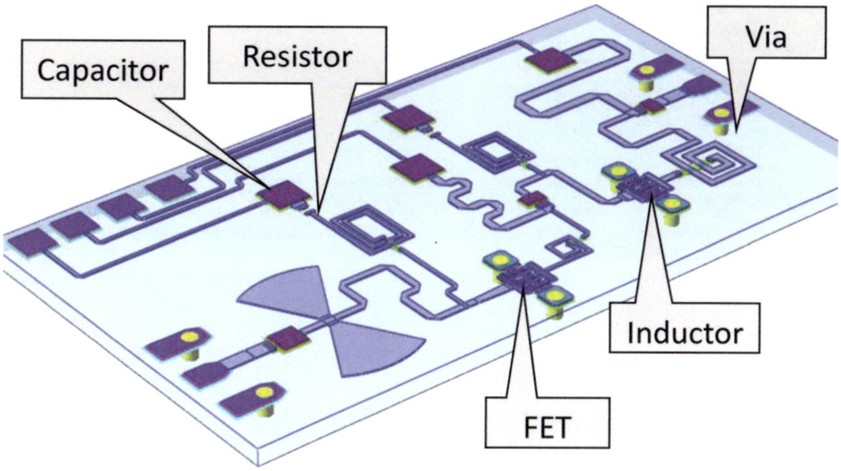

Figure 2.
Layout of MMIC chip with passive and active components.

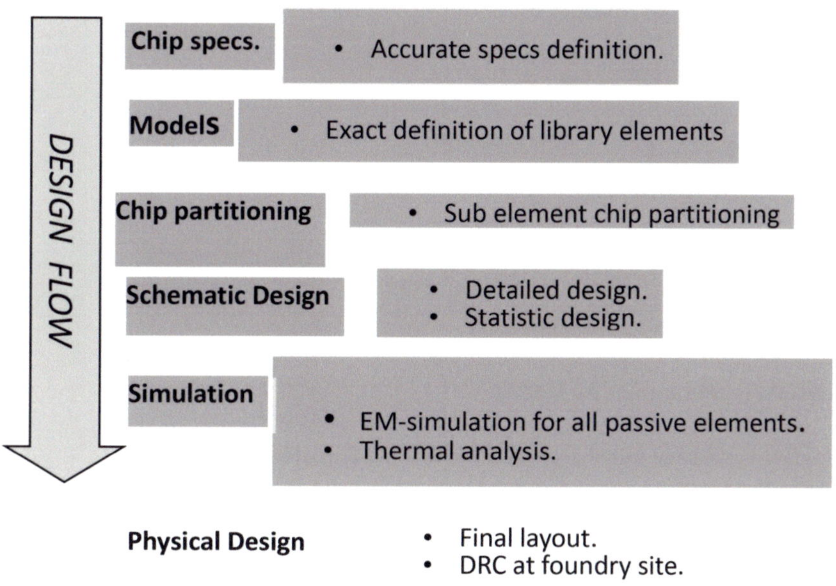

Figure 3.
MMIC design process.

2.2.2 MMIC technologies processes

- 0.25micron GaAs PHEMT amplifiers for power applications at 12GHz to 18GHz.
- 0.15micron GaAs PHEMT for applications at 18GHz to 40GHz.
- GaAs PIN process for switching applications with low power loss.
- HBT, SiGe, InP, GaN, RFMEMS, RFCMOS are new Ka band process

Wafer size may be "3, 5" or 6″. **Figure 4** presents a chip layout located on GaAs WAFER.

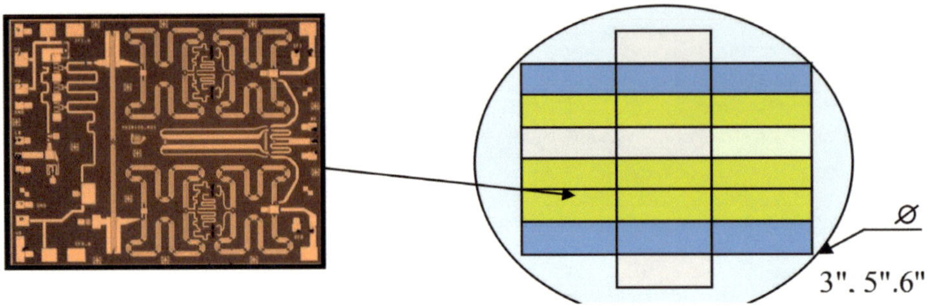

Figure 4.
GaAs WAFER layout and assembly.

2.2.3 Types of components designed using MMIC technology

- Amplifiers - LNA, Power amplifiers, wideband power amplifiers, Distributed TWA
- Mixers - balanced, Star, sub-harmonic
- Switches - PIN, PHEMT, T/R matrix
- Frequency multipliers - active, passive
- FET- Field Effect Transistor
- HEMT- High Electron mobility transistor
- PHEMT- pseudo-morphic HEMT
- MHEMT- metamorphic HEMT
- D-HBT – Double hetero-structure bipolar transistor
- CMOS- Complementary metal-oxide semi-conductor
- BJT- Bipolar Junction transistor
- Modulators - QPSK, QAM (PIN, PHEMT)
- Multifunction - RX chip, TX chip, Switched Amp chip, LO chain

Table 1 presents types of devices fabricated by using MMIC Technology.

Diode	**BJT**	**FET**	**Material**
Schotky GaAs	HBT GaAs D-HBT InP	PHEMT GaAs HEMT InP MHEMT GaAs HEMT GaN	III-V-based
	HBT SiGe	CMOS	Silicon

Table 1.
Materials used in MMIC technology.

2.3 Advantages of GaAs versus silicon in MMIC design

Traditionally low frequency MMICs are produced on silicon substrate. Production costs of MMICs on silicon substrate are cheaper. High frequency MMICs are produced on gallium arsenide (GaAs), a III-V compound semiconductor. MMICs are compact and have low volume and area (from around 1 mm^2 to 10 mm^2). MMICs can be produced in low-cost mass production. The electronic properties of GaAs are significantly better than those of silicon. GaAs has a higher electron mobility and higher saturated electron velocity than silicon. These properties allow transistors produced on GaAs to operate at frequencies higher than 0.3THz. In comparison to silicon devices, GaAs chips are less sensitive to heat because their higher bandgap. Noise of GaAs modules at high frequencies is lower considerably than the noise of silicon modules because of lower resistive device parasitic and higher carrier mobility. These features make GaAs chips and modules attractive to smartphones, cellular phones, medical communication systems, radars, and high frequency phased arrays. Gunn diodes are produced on GaAs substrate to generate RF signals. GaAs devices can be used to emit light efficiently since they have a direct band gap. Silicon devices are very poor at emitting light due to their indirect band-gap. Recent advances may make silicon lasers and LEDs possible. Si LEDs cannot emit visible light and rather work in IR range due to theirs lower bandgap. However, GaAs LEDs may function in visible red light. GaAs substrate is a good choice in high power applications for space electronics devices and optical applications. Silicon is a cheaper substrate than GaAs substrate. Silicon crystal has a significantly mechanically stable structure. Silicon can be grown to very large diameter units. Silicon modules have very high yields. Silicon modules are very attractive for design and production of very large ICs due to good thermal properties of silicon which enable very dense packing of transistors. Silicon dioxide is one of the best insulators, this is a major advantage of Silicon. Silicon dioxide can easily be used in silicon devices. Silicon dioxide layers are adherent to the underlying Silicon layer. GaAs does not have does not have stable oxide does not form a stable adherent insulating layer. An important advantage of silicon over GaAs is the higher hole mobility of silicon which allows the production of higher-speed P-channel field effect transistors. These transistors are required for CMOS logic. GaAs transistors lack a fast CMOS structure. So, GaAs logic circuits have much higher power consumption, GaAs logic cannot compete with silicon logic modules. Silicon technology has lower production cost compared with GaAs devices, contributing to a cheaper Silicon IC. Silicon wafer diameters are typically, 20 cm or 28 cm. GaAs wafer diameters are 10 cm to 15 cm. Other, Indium Phosphide (InP) III-V technologies, offers better properties than GaAs in terms of higher cutoff frequency, gain, and noise figure. InP devices are more expensive than silicon and GaAs modules. InP wafer sizes are smaller than GaAs wafers and are more fragile. Silicon Germanium technology offers higher speed transistors than conventional Silicon devices with similar cost expenses. Gallium Nitride, GaN, is used to produce power amplifiers MMICs. GaN transistors can work at much higher voltages and function at much higher temperatures than GaAs transistors, they are used to produce power amplifiers at high frequencies. Properties of dielectric substrates used in MMIC technology are given in **Table 2**.

2.3.1 Semiconductor in MMIC devices

Si CMOS MMIC modules are low power and low-cost devices. Si CMOS MMIC modules may operate in frequencies lower than 0.2THz. SiGe MMIC devices are used as medium power high gain devices. SiGe MMIC modules may operate in frequencies lower than 0.2THz. InP HBTs modules may operate in frequencies lower

InP	GaAs	Si or Sapphire	Si	Property
14	12.9	11.6	11.7	Dielectric constant
10^7	10^7–10^9	>10^{14}	10^3–10^5	Resistivity Ω/cm
3000	4300	700	700	Mobility(cm^2/v-s)
4.8	5.3	3.9	2.3	Density (gr/cm^3)
1.9×10^7	1.3×10^7	9×10^6	9×10^6	Saturation velocity(cm/s)

Table 2.
Comparison of material properties in MMIC technology.

than 0.4THz. InP HBT modules are used as high frequency medium power high gain devices. InP HEMT modules may operate in frequencies lower than 0.6THz. InP HEMT devices are used as high frequency medium power high gain devices. Properties of MMIC technologies are presented in **Table 3**. PHEMT, 0.15micron, on GaAs substrate is shown in **Figure 5**.

GaN HEMT	InP HEMT	InP HBT	SiGe HBT	Si CMOS	
<0.2THz	<600GHz	<0.4THz	<0.2THz	<0.2THz	High frequency
<0.2THz	<0.68THz	<0.33THz	<0.25THz	<0.2THz	MMICs
High	Medium	Medium	Medium	Low	P-Out
Low	Low	High	High	Low	Gain
Low	Low	High	High	High	Noise
Low	Low	Medium	High	High	Yield
No	No	Yes	Yes	Yes	Mixed signal
High	High	Low	Low	High	1/f noise
>20 V	-2 V	-4 V	-2 V	-1 V	Breakdown Voltage

Table 3.
Summary of semiconductor for MMIC technology.

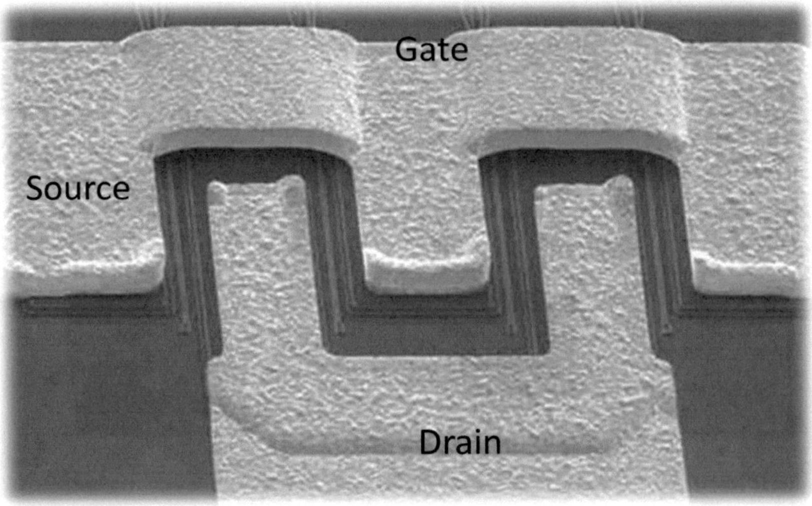

Figure 5.
Photo of 0.15micron PHEMT MMIC on GaAs substrate.

2.4 Generation of microwave signals in microwave and mm wave

RF Signals can be excited in vacuum-tube based devices and in solid-state oscillators. Solid state modules are produced on semiconductors such as silicon, SiGe, gallium arsenide, GaN and include bipolar junction transistors (BJTs), field-effect transistors (FETs), Gunn diodes, and IMPATT diodes. RF variations of BJTs include the heterojunction bipolar transistor (HBT), and RF variants of FETs include the MESFET, the HEMT, and LDMOS transistor. RF waves can be generated and processed using MIC circuits and MMICs. Traditionally these modules are produced on gallium arsenide wafers. However, silicon germanium and heavy-dope silicon are recently used to produce high power modules. Vacuum tube high power modules operate on the ballistic motion of electrons in a vacuum under the influence of controlling magnetic or electric fields. These devices work in the density modulated mode, rather than the current modulated mode. They work on clumps of electrons flying ballistically through them. These devices include traveling wave tube, klystron, magnetron, and gyrotron.

2.5 MMIC modules and applications

A wide band Ka band power Amplifier is shown in **Figure 6**. The input power is divided by using a splitter. The microwave signal is amplified by power amplifiers and combined by a power combiner to get the desired output power.

A Ka Band wideband non reflective MMIC Pin diode SPDT is shown in **Figure** 7. MMIC process cost per the MMIC area is given in **Table 4**.

MMIC Applications

- Ka band satellite communication.
- 60GHz wireless communication.
- Automotive Radars
- Imaging in security
- Gbit WLAN

Figure 6.
Wide band MMIC mm-wave power amplifier.

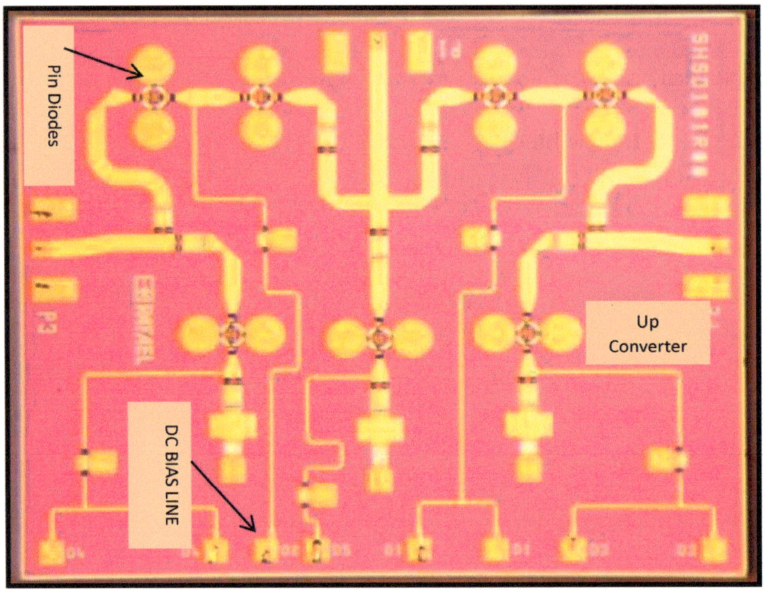

Figure 7.
Wideband Ka band non reflective SPDT.

InP HEMT	GaAs HEMT	SiGe HBT	Si CMOS	
10	1–2	0.1–0.5	0.01	Chip Cost($/mm^2)
0.0135	0.0135	0.135	1.35	Mask cost (M$/mask set)

Table 4.
MMIC COST.

3. MEMS technology

Micro-Electro-Mechanical Systems (MEMS) technology integrate sensors, actuators, mechanical elements, and electronics on the same silicon substrate using microfabrication technology. MEMS modules replace connectorized devices, actuators, sensors, and antennas with micron scale similar devices that can be produced in mass production by a production process using integrated circuits and photolithography technology. MEMS devices reduce size, cost, weight, and power consumption while improving properties, production cost and yield, volume, and functionality significantly. The dimensions of MEMS modules may vary from several millimeters to around one micron. MEMS modules may vary from very simple structures to structures with moving elements. There are complex MEMS electromechanical systems with several moving elements controlled by integrated microelectronics. During the last thirty years MEMS designers, developers and researchers have produced an extremely large number of MEMS sensors for several sensing applications. For example, pressure sensing, heartbeat, temperature sensing, inertial forces, chemical species, magnetic fields, radiation detection and movement detection. Usually, these MEMS sensors have better performances exceeding those of conventional sensors and devices. The electronics components are produced using IC process. The micromechanical components are produced by using compatible "micromachining" processes. These processes, by using masks, etch away parts of the silicon wafer or add new structural layers to form the electromechanical and mechanical modules.

The real potential of MEMS may be fulfilled when these miniaturized sensors, actuators, and other components can all be merged onto a common silicon substrate along with integrated circuits, microelectronic ICs. The electronic components are fabricated using integrated circuit (IC) process sequences (such as CMOS, Bipolar, or BICMOS processes). The micromechanical components are fabricated using compatible "micromachining" processes that selectively etch away parts of the silicon wafer or add new structural layers to form the mechanical and electromechanical devices.

3.1 MEMS technology features and advantages

- Insertion loss lower than <0.1 dB
- Isolation lower than -50 dB
- High Linearity compared to conventional devices
- High Q compared to conventional devices
- Low volume and compact
- High power handling compared to conventional devices
- Low power consumption compared to conventional devices
- Low-cost and high-volume production compared to conventional devices

3.2 MEMS technology process

Bulk micromachining produces mechanical structures in the silicon substrate by using etching masks. Bulk micro-machined module is shown in **Figure 8**.

Surface Micromachining produce mechanical structures above the substrate surface by using sacrificial layer. Surface micro-machined module is presented in **Figure 9**. In Bulk micromachining technology silicon is machined using etching processes. Surface micromachining uses layers deposition on the substrate to produce a structural layer.

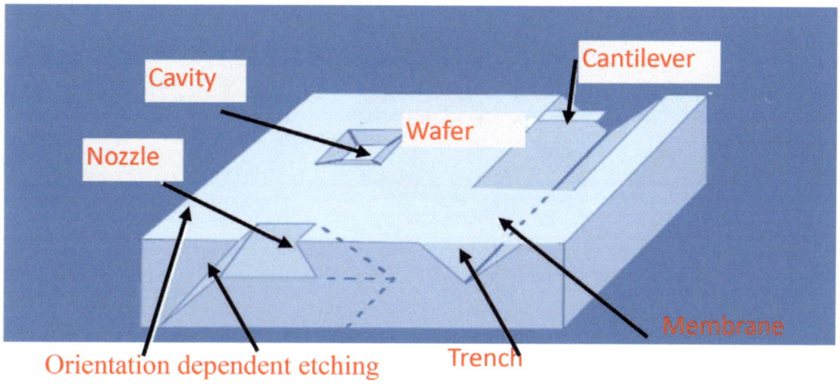

Figure 8.
MEMS bulk micromachining technology.

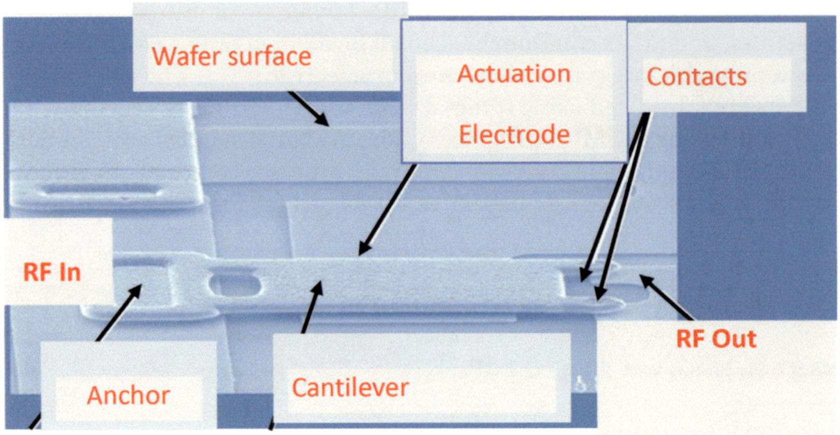

Figure 9.
MEMS surface micromachining technology.

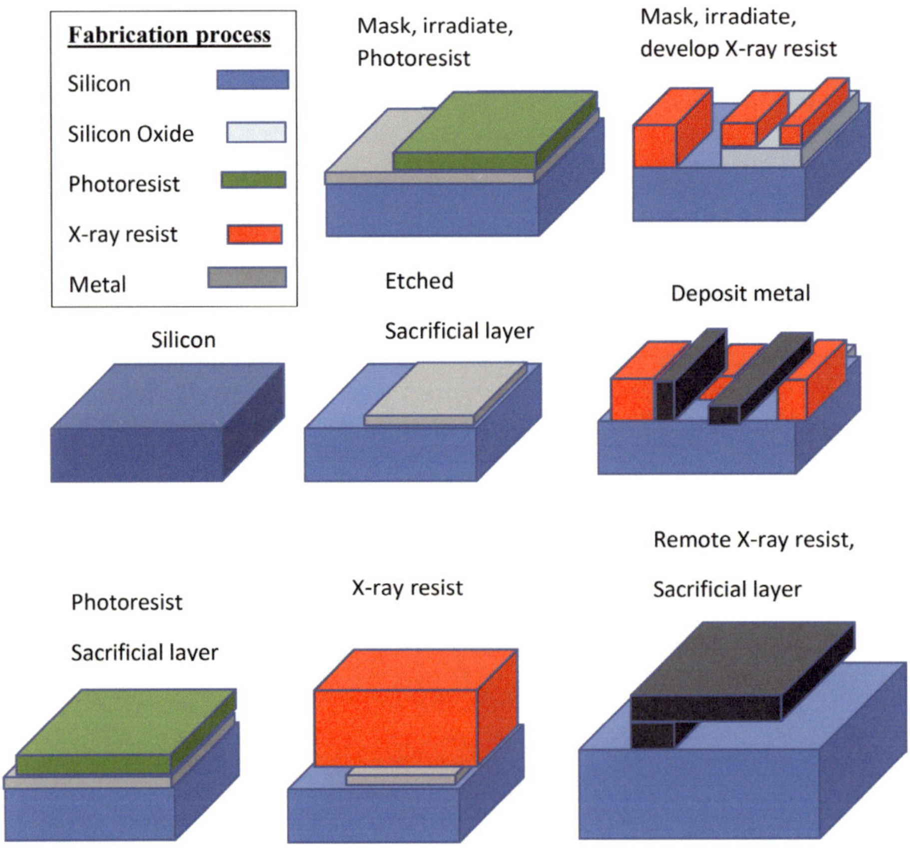

Figure 10.
MEMS *fabrication process.*

Surface micromachining process does not depend on the substrate used. It can be part of other production processes that modify the substrate. For example, fabrication of MEMS on a substrate with embedded control devices, in which MEMS process is integrated with integrated circuit technology.

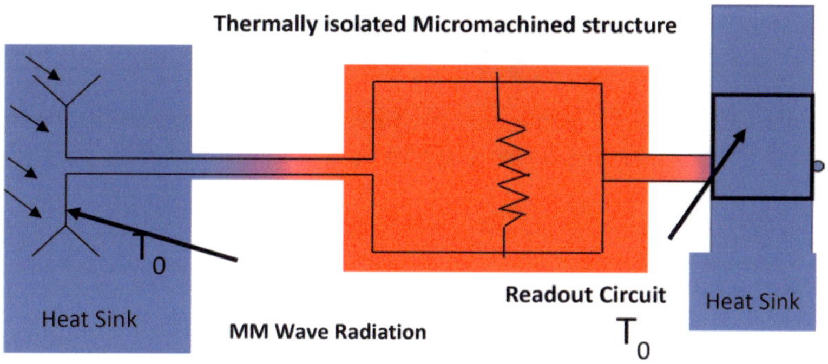

Figure 11.
MEMS bolometer coupled antenna array.

This process is used to manufacture a wide spectrum of MEMS modules for several applications. Bulk micromachining is a subtractive fabrication process, that converts the substrate, into the mechanical parts of the MEMS module. MEMS modules can be designed by using electromagnetic software such as ADS, HFSS, and CST. The design outcomes layers masks, layout that are used to produce the MEMS module. MEMS production process is shown in **Figure 10**. In **Figure 11** the block diagram of a MEMS bolometer coupled antenna array is presented. Packaging of MEMS modules may be more complicated. However, higher devices are easier to produce when compared to surface micromachining. Applications of RF MEMS technology:

- Tunable microwave MEMS passive elements such as inductors and filters
- MEMS Switching matrix with low loss
- MEMS antenna arrays coupled to bolometer for detection arrays
- MEMS 90GHz Detection Arrays

3.3 MEMS components

MEMS components are categorized in one of several applications. Such as:

1. **MEMS The goal of MEMS sensors** is to sense changes and interact with the environments. MEMS sensors include sensors that detect change in temperature, sensors that detect chemical changes, movement sensors, optical sensors, radiation sensors and inertia sensors. MEMS sensors are useful due to their low-cost and small volume.

2. **Microwave MEMS** are modules used to transmit microwave signals, to switch or to filter RF signals. **Microwave MEMS** include tunable capacitors, switches, filters, antennas, and phase shifters.

3. **Optical MEMS** are compact modules that amplify, direct, reflect, and filter light such as optical reflectors and switches.

4. Thermal or electrostatic actuators provide power to other components or devices.

5. **Biological MEMS** are modules that, interact with biological tissue. These modules interact with biological cells, medical reagents, proteins, and other

biological tissues. These devices may be employed for drug delivery or other medical missions.

6. **Microfluidic MEMS** are low-cost modules that interact with fluid environments. Modules such as pumps and valves. Devices that move, mix, and eject and are compact.

4. Computer aided design, CAD, commercial software

In the last decade, several electromagnetic commercial software was developed, see [16–19]. The most popular software that are used in the design and development of wearable systems and antennas will be presented in this section.

4.1 High frequency structure simulator, HFSS, software

ANSYS HFSS RF simulation software is a full wave electromagnetic software. HFSS is used to simulate RF modules and antennas [17]. For example, printed antennas, waveguides and other transmission lines, connectors, antenna arrays, phased arrays, microwave devices, digital circuits, filters, MIC and MMIC packages and other RF devices. ANSYS HFSS is used to develop RF, high-speed RF systems, seekers, radar systems, microwave devices, RF satellites modules, internet of things (IoT) products and other high-speed microwave and digital devices. For more information see, https://www.ansys.com/products/electronics/ansys-hfss.

HFSS employs several solvers to give the RF engineer deep insight into all the 3D electromagnetic problems. Through integration with ANSYS thermal, structural and fluid dynamics tools, HFSS provides a multi-physics analysis of RF products, ensuring their thermal and structural reliability. The ANSYS HFSS simulation suite consists of a comprehensive set of solvers to address diverse electromagnetic problems. Its automatic adaptive mesh refinement lets the designer to focus on the design instead of spending time determining and creating the best mesh.

4.1.1 HFSS electromagnetic solvers

ANSYS HFSS employs full wave 3D finite element method, momentum (MoM) analysis and the SBR ultra-large-scale asymptotic method of shooting and bouncing rays with advanced diffraction and creeping wave physics for enhanced accuracy (SBR+). The ANSYS HFSS software package has the following solvers to solve RF problems:

HFSS Solvers

- Solver in the frequency domain
- Solver in the time domain
- Solver using integral equations
- Solver that uses hybrid technologies

HFSS SBR+

- Solver that uses physical optics
- Solver that uses shooting and bouncing Ray

- Solver that uses physical theory of diffraction
- Solver that uses Creeping Wave
- Solver that uses uniform theory of diffraction

Each HFSS may solve a specific module, environment, RF devices or application.

HFSS RF Solver Features

- Communication system link budget calculations
- Wireless propagation solver
- Microwave and antennas solver
- Automated diagnostics simulation
- Microwave component libraries
- Multi-fidelity RF models
- Coupling models and analysis

Types of Circuit Simulation

- Linear circuit analysis
- DC circuit analysis and simulation
- Transient circuit analysis and simulation
- RF system multitone harmonic balance analysis

For more information see, https://www.ansys.com/products/electronics/ansys-hfss.

4.2 Advance design system, ADS

ADS is an electronic design automation software system. It offers complete design integration to designers of products such as cellular and portable phones, pagers, wireless networks, and radar and satellite communications systems [16].

ADS support communication system and RF design engineers to develop all types of RF designs, from RF and microwave modules and printed antennas to integrated MMICs for communication, medical, IOT and aerospace defense applications.

With a complete set of simulation technologies ranging from frequency and time domain circuit simulation to electromagnetic field simulation. ADS let designers fully characterize and optimize designs. Such as Harmonic Balance, Circuit Envelope, Transient Convolution, Ptolemy, X-parameter, Momentum, and 3D EM simulators (including both FEM and FDTD solvers).

ADS Simulation and Design Major Features

- Development and design of communication and RF systems
- Analysis results presentation, S parameters data and plots
- S-parameters simulation, DC, and small signal AC simulation
- Simulation of RF devices, modules, and systems
- Yield statistical simulation including components tolerances simulation to get accurate design
- Optimization of component parameters to get accurate design and high yield
- Development and design of several types of filters
- Development and design of passive devices, S-parameters simulation
- Developer of customized design guides
- ADS software libraries such as microwave component and systems, passive and active components
- Design and development of MIC and MMIC modules
- Electromagnetic simulation, radiation simulation
  ADS Suite Features
- **Design tools** - Schematic graphical friendly user interface. Circuit schematic entry and simulation setup.
- **Computation results presentation** – Display graphically and tables of analysis results.
- **Connection Manager** – Allow to read data from data files downloads. Control measurement instruments.
  Simulators
- Frequency-domain linear RF modules simulator.
- Optimization simulation, Yield simulation, Yield Optimization, experiments design, and Presentation of Statistical yield histograms.
- **Electromagnetic Momentum Simulator** - Electromagnetic Momentum, EM, Simulator is a frequency domain simulation CAD tool. EM combines layout editing tools with electromagnetic simulation and S-parameters simulation to get accurate RF design. EM analysis suite includes schematic tools, simulation results presentation, EM simulator, and multilayer layout editor.
- **Microwave and Communication System Simulation** – Modular analysis of microwave and communication systems. Analysis of each unit and component in the system.

- **Models of Passive Components** - Models for capacitors, inductors, transformers, couplers, conductors, vias, and crystals.
- **Models for Multilayer Design** - Coupled lines models for multilayer modules such as MIC PCB, MMIC modules, and devices packaging.
- **Models for Active Microwave Devices** – Amplifiers, modulators and demodulators, mixers, and switches.
- **Design Guide for Passive Devices** - Simulation and optimization of microstrip circuits such as couplers, branch-line couples, dividers, coupled line filters, microstrip matching circuits and lumped-element circuits.
- **Passive Filters Simulator** - Simulation and optimization of passive filters.

4.2.1 FEM simulator

The FEM simulator has a full-wave three dimensions electromagnetic simulation capabilities, based on the Finite Element Method (FEM). The RF-Pro User Interface (UI), that comes with this FEM element, makes setting up RF circuit co-simulation in Advanced Design System (ADS) fast with no errors for the design of multi-technology RF modules that integrate RFIC, MMIC, package and PCB. It also automates the extraction of nets and components for EM simulation without modifying the layout. The FEM simulator enable to simulate 3D structures such as connectors, wire-bonds and packaging with circuit and system components. It is important especially for RF module designs where 3D interconnects and packaging must be simulated along with the circuit. For more details see in, https://www.keysight.com/en/pc-1297113/advanced-design-system-ads?nid=-34346.0&cc=IL&lc=eng.

The frequency domain simulation can be used in the Electromagnetic Professional (EM-Pro) software, in the 3D EM platform and in ADS.

4.3 CST software

CST is a **3D** electromagnetic **simulation** software for developing, designing, analyzing, and optimizing RF components and systems [18]. Information about CST is presented in, https://www.3ds.com/products-services/simulia/products/cst-studio-suite/.

CST Electromagnetic field solvers gives engineers the flexibility to RF systems that consists of many components and modules. CST SIMULA software allows electromagnetic simulation to accompany the design process from the earliest design stages to the final development milestone. CST EM analysis include design of couplers, dividers, antennas, filters, MEMS, electromagnetic compatibility simulation (EMC/EMI), simulation near human body, and thermal effects in transmitters. SAM a System Assembly and Modeling tool provides simple environment simulation software RF systems. The SAM suite may be employed for simulating, analyzing, and optimizing RF devices that consists of several components. The simulation products are physical quantities such as voltages, currents, fields, and S-parameters. SAM helps designers to compare the results of different solvers within one analysis project and perform post-processing simulation. For example, using the results of electromagnetic simulation to analyze thermal effects, then structural thermal effects, and other EM simulation to analyze detuning. This analysis process helps to reduce the calculation effort required to analyze a complex device accurately. The

CST software is a 3D modeling schematic layout tool, with electromagnetic solvers and post-processing simulations.

CST Solvers

- Transient solver
- TLM solver
- Frequency domain solver
- Eigenmode solver
- Resonant solver
- Integral Equation Solver
- Asymptotic Solver

CST Products

- CST EM analysis is used to design several RF devices and systems. Such as couplers, dividers, antennas, filters, MEMS, electromagnetic compatibility simulation (EMC/EMI), simulation near human body, SAR problems, and thermal effects in transmitters.
- CST Simulation products include static, stationary, high and low frequency analysis of RF systems, and components with movement of charged particles.
- CST software is used to analyze multilayer structures, transmission lines, EMC design, small antennas, antenna arrays, packaging, LTCC devices, inductors, capacitors, waveguide devices, actuators, plasma sources, optical devices, sensors, recording units, and electromagnetic brakes.

4.4 Microwave office, AWR

Microwave Office design suite provides a flexible RF/microwave design tool [19].

Built on AWR high-frequency design platform with its open design environment and advanced unified data model. AWR operates with Visual System Simulator, VSS, system design, AXIEM and Analyst EM simulation software. The NI AWR Design suite provide a complete microwave devices solver, system solver, and EM simulation.

Microwave Office AXIEM electromagnetic (EM) software is an EM analysis.

The AXIEM product was developed specifically for three-dimensional (3D) planar applications such as RF PCBs and modules, LTCC, MMIC, and RFIC designs.

The APLAC simulator offers multi-level analyses which includes:

- DC operation point
- Linear frequency domain
- Time domain
- Harmonic balance

- Phase noise
- Linear/non-linear noise including AC noise contributors, temperature
- Yield predictions and optimization

More information can be found in https://www.awr.com/serve/microwave-office-brochure-1.

AWR provide accurate simulation tool and offers RF devices analysis and optimization. AWR provides a linear and nonlinear time and frequency domain simulation needed to characterize and optimize RF modules. AWR major features are listed below.

- **Microwave and Communication System Simulation** – Modular analysis of microwave and communication systems. Analysis of each unit and component in the system.
- **Models of Passive Components** - Models for capacitors, inductors, transformers, couplers, conductors, and SMT components.
- **Models for Multilayer Design** - Multilayer modules such as MIC PCB, MMIC modules, and devices packaging.
- **Active Microwave Devices Simulation** – Linear and nonlinear simulation, Amplifiers, modulators and demodulators, mixers, and switches.
- **Passive Devices** Simulation - Simulation and optimization of microstrip circuits such as couplers, branch-line couples, dividers, coupled line filters, microstrip matching circuits and lumped-element circuits.
- **Passive Filters Simulator** - Simulation and optimization of passive filters.

AWR Capabilities– The AWR user interface provides project management and design tools for RF devices. The designers can build a device layout model from the software component library. The library supports tuning and optimization simulation.

- **Computation results presentation** – Display graphically and tables of analysis results.

Simulation APLAC – This robust harmonic-balance (HB) simulator provides linear and nonlinear circuit analysis with powerful multi-rate HB, transient-assisted HB, and time variant (circuit envelope) analysis, supporting large-scale and highly nonlinear RF/microwave circuits.

Planar EM – AXIEM provides accurate characterization, simulation, optimization of passive devices, planar transmission lines, printed antennas, and printed arrays.

3D EM – 3D finite element solver provides fast and accurate electromagnetic analysis of multilayer elements such as vias and bonds. The high frequency full wave analysis helps in developing RF modules from the beginning of the design up to the final electromagnetic verification.

AWR for MMIC Modules Design

- Linear and nonlinear MMIC Modules design with frequency-domain and time-domain simulation.

- MMIC module layout and production files and masks, GDSII export.
- MMIC module electromagnetic simulation and optimization from layout or schematic using commercial electromagnetic solvers.
- MMIC module design and layout rule check versus schematic.

5. Ultra-wideband compact integrated modules

UWB integrated modules of 18 to 40GHz Direction Finding system are shown in **Figures 12** and **13**. A photo of a compact UWB frontend is shown in **Figure 12**. The frontend module consists of a limiter and a wideband 18 to 40GHz LNA. The low noise amplifier LMA406 is A Filtronic MMIC LNA. The LNA gain is around 11 ± 1 dB with 4 ± 0.5 dB noise figure and 13 ± 1 dBm saturated output power. The LNA dimensions are 1.45 × 1.1 × 0.1 mm. A wide band PHEMT MMIC SPDT is used,

Figure 12.
18 to 40GHz frontend module.

Figure 13.
A photo of UWB switched filter Bank.

AMMC-2008. The SPDT losses are lower than 2 dB. The isolation between the SPDT input port to the output ports is lower than -25 dB. The SPDT dimensions are 1 × 0.7 × 0.1 mm. The frontend electrical characteristics was evaluated by using ADS software.

Figure 13 presents a photo of the compact Switched Filter Bank SFB unit. The SFB Module consists of three side coupled microstrip filters. Each filter consists of nine sections. The filters are printed on a 5mil alumina substrate. A one to two dB attenuators connect the filters input and output ports to wide band MMIC switches. The attenuators are used to adjust each channel losses to the average required level. The module losses are adjusted to be higher in the low frequencies and lower in the high frequencies. AWR and ADS software were used to optimize the filter dimensions and structure. The SFB losses at high frequencies are around 9 dB and at the low frequencies the losses are around 10.5 dB.

Author details

Albert Sabban
Senior Researcher, Lecturer and Electromagnetic Consultant, Ort Braude College, Karmiel, Israel

*Address all correspondence to: sabban@netvision.net.il

References

[1] J. Rogers, C. Plett "Radio frequency Integrated Circuit Design", Artech House, 2003.

[2] N. Maluf, K. Williams, "An Introduction to Microelectromechanical System Engineering", Artech House, 2004.

[3] Albert Sabban (2016). "Wideband RF Technologies and Antenna in Microwave Frequencies", Wiley Sons, July 2016, USA.

[4] Albert Sabban Editor and Author (2020). "Wearable Systems and Antennas Technologies for 5G, IOT and Medical Systems" CRC PRESS, TAYLOR & FRANCIS GROUP, December 2020. ISBN 9780367409135.

[5] Albert Sabban (2017). "Novel Wearable Antennas for Communication and Medical Systems, TAYLOR & FRANCIS GROUP, October 2017.

[6] Albert Sabban (2015). "Low visibility Antennas for communication systems", TAYLOR & FRANCIS GROUP, 2015, USA.

[7] S. A. Mass, "Nonlinear Microwave and RF Circuits", Artech House, 1997.

[8] Abdeen Hebat-Allah Yehia; Yuan Shuai; Schumacher Hermann; Ziegler Volker; Meusling Askold; Feldle Peter." 10 to 40 GHz Superheterodyne Receiver Frontend in 0.13 µm SiGe BiCMOS Technology". Frequenz, Volume 71, pp.151-160. March 2017.

[9] A. Sabban, "Microstrip Antenna Arrays", 2011, Microstrip Antennas, Nasimuddin Nasimuddin (Ed.), ISBN: 978-953-307-247-0, InTech, pp. 361-384,

[10] A. Sabban, "Applications of MM Wave Microstrip Antenna Arrays" ISSSE 2007 Conference, Montreal Canada, August 2007.

[11] Reuven Shavit, Albert Sabban, Michael Sigalov, Avihai Lahman, Zeev Iluz, Naftali Chayat and Solon Spiegel "Microwave Engineering Research Activity in Israel", IEEE Microwave Magazine, May 2018, 19(3):129-135. DOI: 10.1109/MMM.2018.2802281

[12] G.P. Gauthier, G.P. Raskin, G.M. Rebiez., P.B. Kathei," A 94GHz Micro-machined Aperture- Coupled Microstrip Antenna", I.E.E.E Trans. on Antenna and Propagation Vol. 47, No. 12, pp. 1761-1766, December 1999.

[13] G. de Lange et. al., "A 3*3 mm-wave micro-Machined Imaging array with sis Mixers", Appl. Phys. Lett. 75 (6), pp. 868-870, 1999.

[14] A. Rahman et. al., "Micro-machined room Temperature micro bolometers for MM-Wave Detection", Appl. Phys. Lett. 68 (14), pp. 2020-2022, 1996.

[15] A. Sabban and K.C. Gupta, "Characterization of Radiation Loss from Microstrip Discontinuities Using a Multiport Network Modeling Approach", I.E.E.E Trans. on M.T.T, Vol. 39, No. 4, April 1991, pp. 705-712.

[16] Keysight software, http://www.keysight.com/en/pc-1297113/advanced-design-system-ads?cc=IL&lc=eng

[17] ANSYS HFSS, https://www.ansys.com/products/electronics/ansys-hfss.

[18] CST Software, https://www.3ds.com/products-services/simulia/products/cst-studio-suite/

[19] Microwave office AWR, https://www.awr.com/serve/microwave-office-brochure-1.

Section 2

Ultra-Wideband Systems and RF Modules

Chapter 2

Ultra-Wideband MM Wave System and RF Modules

Albert Sabban

Abstract

Compact wideband RF modules are crucial in mm-wave direction finding systems, radars, seekers, and communication systems. This chapter discusses new integrated wideband mm-wave RF modules. It also discusses the design and development of a compact wideband (18–40 GHz) frontend and a wideband (18–40 GHz) switch bank filter (SBF). The frontend electrical specifications determine the system signal-to-noise ratio and the system dynamic range. This chapter presents a low-cost integrated 18–40 GHz wideband compact frontend with a 47 dBm high power limiter. The frontend consists of two channels: a high gain and low gain channel. Wideband MMIC switches are employed to select the required channel. The gain of the high gain channel is around 27 dB with ±1 dB flatness. The noise figure of the module is around 9 dB. This chapter also presents a low-cost, integrated, 18–40 GHz wideband compact SFB module. The wideband SFB consists of three wideband side-coupled microstrip filters. The SFB MIMIC switches operate in the 18 to 40 GHz frequency range and are used to select the required filter. The insertion loss of each filter section is less than 11.5 dB ±1.5 dB. The novelty of this research is the development of compact, integrated wideband mm-wave RF modules for direction finding and communication systems.

Keywords: MM Wave Direction Finding System, Frontend, Filters, Wideband communication systems

1. Introduction

Design of wideband RF modules, filters, and antennas are presented in [1–12]. Wideband RF technologies such as MIC, MIMIC, and MEMS are presented in [1–7]. Wideband RF modules are crucial in the development of direction finding (DF) systems. A fully integrated 10–40 GHz superheterodyne receiver frontend using a 40–46 GHz IF is presented in [8]. Wideband RF technologies are used to develop wideband RF modules such as frontends, SFBs, and receiving and transmitting channels, as presented in [1–15]. A DF system measures the direction from which a received signal was transmitted. Radio DF is used in the navigation of ships, air-crafts, vehicles, and missiles to locate emergency transmitters for search and rescue, to locate illegal or interfering transmitters, and to track wildlife. The transmitted signal direction may be found by combining the direction information from two or more suitably spaced receivers by using triangulation. Triangulation is the process of determining the location of a point by forming triangles. Triangulation involves only angle measurements. A DF system provides the ability to locate the position of an enemy transmitter.

The demand for wide bandwidth makes the Ka band attractive for future commercial communication and the radar industry. Front-end modules, filter banks, and coupler modules are important units in radar, seekers, and communication systems. This chapter presents several wideband mm-wave RF modules. It also discusses the design and development of a compact wideband (18–40 GHz) frontend and a wideband (18–40 GHz) SBF. Advanced design system (ADS) full-wave electromagnetic software is used to develop wideband RF modules, as presented in [16].

2. Ultra-wideband direction finding system

Compact wideband RF modules are crucial in mm-wave direction finding, radars, seekers, and communication systems. **Figure 1** shows a compact 18–40 GHz DF system is shown. **Figure 2** presents the block diagram of a wideband mm-wave DF system. The DF system consists of SFB modules, frequency source unit, DF frontend modules, two omnidirectional frontend units, and downconverter units. Wideband RF technologies are used to develop wideband RF modules such as frontends, filters, and receiving and transmitting channels, as presented in [1–5]. Development of wideband filters, microwave components, RF modules, and antennas is widely discussed in the literature (see [1–14]).

2.1 Ultra-wideband compact MM-wave frontend1

Design and development of compact wideband frontend is crucial in development of DF systems. The RF modules and the system are designed using RF, ADS software, and momentum RF software, [16]. **ADS** is an electronic design automation software system. It offers complete design integration to designers of products such as cellular and portable phones, pagers, wireless networks, and radar and satellite communications systems.

ADS support communication systems and RF design engineers to develop all types of RF designs, from RF and microwave modules and printed antennas to

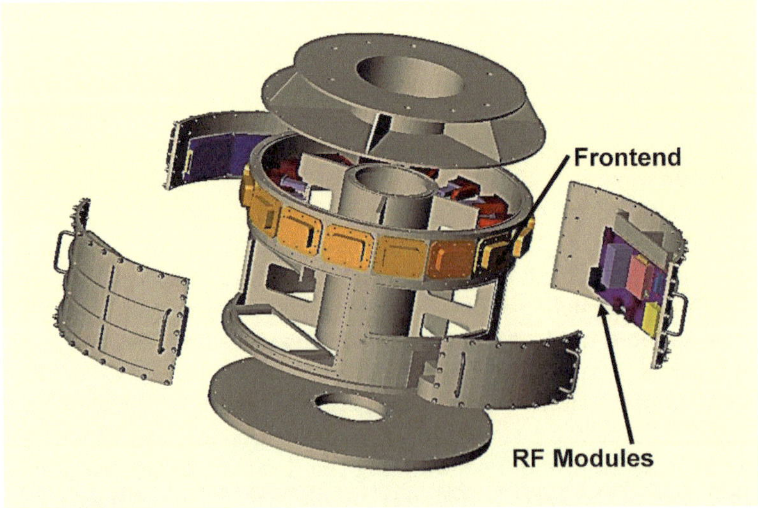

Figure 1.
Ultra-wideband DF system.

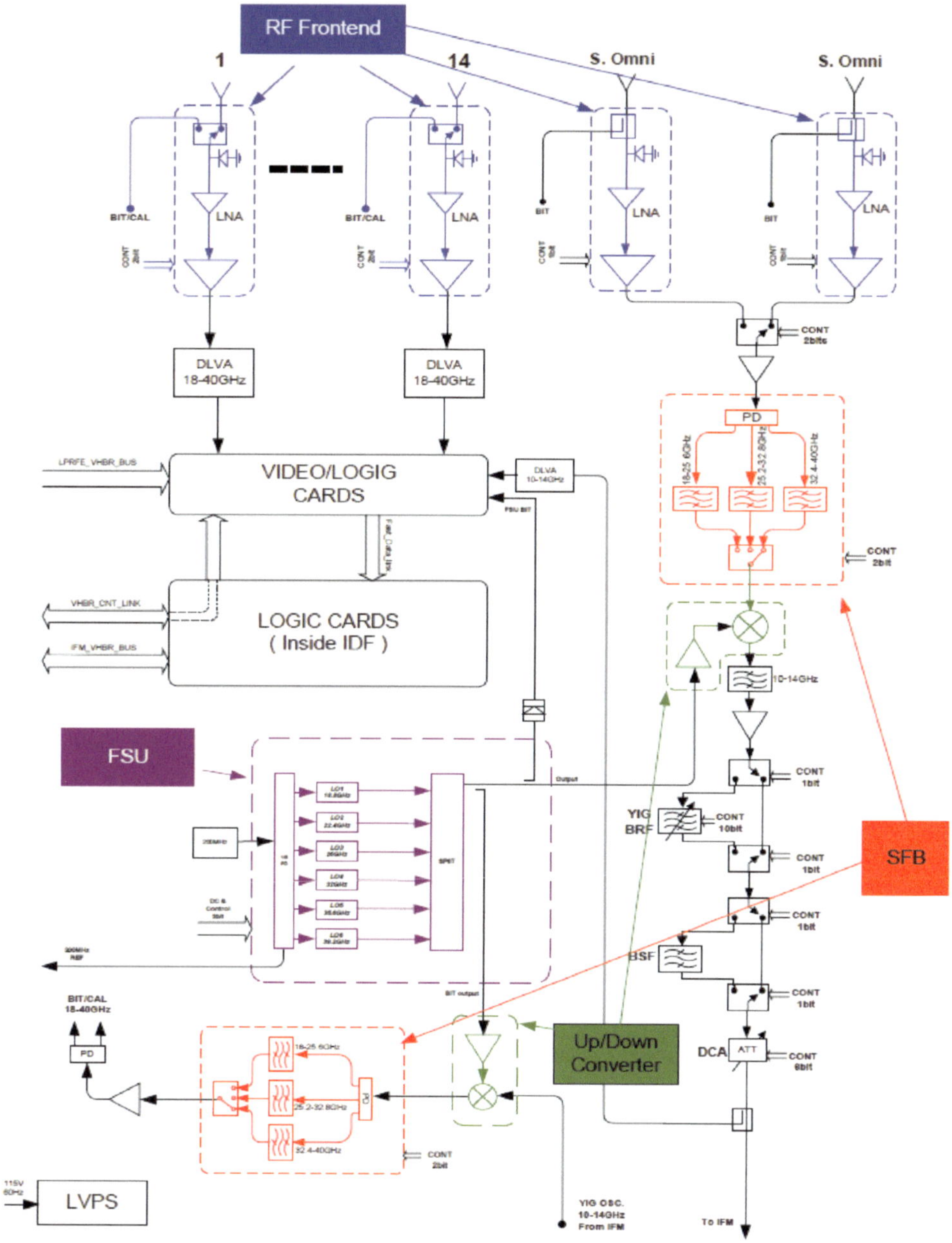

Figure 2.
Block diagram of an 18–40 GHz wideband DF system.

integrated MMICs for communication, medical, and aerospace defense applications as well as the Internet of Things (IoT).

With a complete set of simulation technologies ranging from frequency and time domain circuit simulation to electromagnetic field simulation. ADS let designers fully characterize and optimize designs, such as harmonic balance, circuit envelope, transient convolution, ptolemy, X-parameter, momentum, and 3D EM simulators (including both FEM and FDTD solvers). Measured results approve the computed results and the design process.

2.2 Compact MM-wave frontend requirements

Table 1 lists the frontend electrical specifications. The frontend design presented in this section meets the frontend electrical specifications. **Figure 3** presents the frontend block diagram, and **Table 2** lists the frontend interface connectors.

Physical characteristics - interface connectors

2.3 Wideband front-end design

Figure 3 shows a frontend block diagram. The frontend module consists of a limiter and a wideband 18–40 GHz low noise amplifier (LNA) LMA406. The solid state LMA406 is a low noise PHEMT amplifier that operates from 18 to 40 GHz. The amplifier is a two-stage amplifier.

The amplifier gain is around 12 dB with 4.5 dB noise figure and 14 dBm saturated output power, and − 1 dB gain compression power output of +10 dBm. The LNA dimensions are 1.44 x 1.1 mm. A wideband PHEMT MMIC SPDT, AMMC-2008, was used. AMMC-2008 is a monolithic PHEMT SPDT switch with low insertion loss and high isolation from DC to 50 GHz. For improved reliability and moisture

Parameter	Requirements	Performance
Frequency range	18–40 GHz	Agree
Module gain	24 ± 3 dB, Switched by external control. (Lower than -40 dB for off state)	Agree
Module gain flatness	For any 0.5 GHz BW in 18–40 GHz ±0.5 dB max. For any 4GHz BW in 18–40GHz ±2 dB max. ±3 dB max for the whole range 18–40 GHz	Agree
High gain state noise figure	11 dB over temperature. 10 dB max for 40°C baseplate temperature.	Agree
Input power	−60 dBm to 10 dBm	Agree
Output power	13 dBm saturated −40 dBm to 11 dBm not saturated	Agree
Linearity	12 dBm min. Output at 1 dBc compression point 21 dBm single tone for third intercept point (Ip3) Second harmonic power − 25 dBc max. For 10 dBm output.	Agree
S11	Better than −9 dB	Agree
Input power protection	Input power 1 W CW at 0.1–40 GHz No damage at 1 W CW and 20 W Pulses Test for pulses: PW = 1 usec, PRF = 1KHz	Agree
Voltages	±5 V, +15 V	Agree
Control signals	LVTTL standard “0” = 0–0.8 V; “1” = 2.0–3.3 V	Agree
Switching time	Less than 0.1 usec	Agree
Output spurious, non-harmonic	When it is not, correlative with the input signals −50 dBm max.	Agree
Video leakage	Video leakage signals will be below the RF output level for terminated input	Agree
Module volume	6 x 4 x 2 cm	Agree

Table 1.
Electrical specifications of a wideband, 18–40GHz, DF frontend.

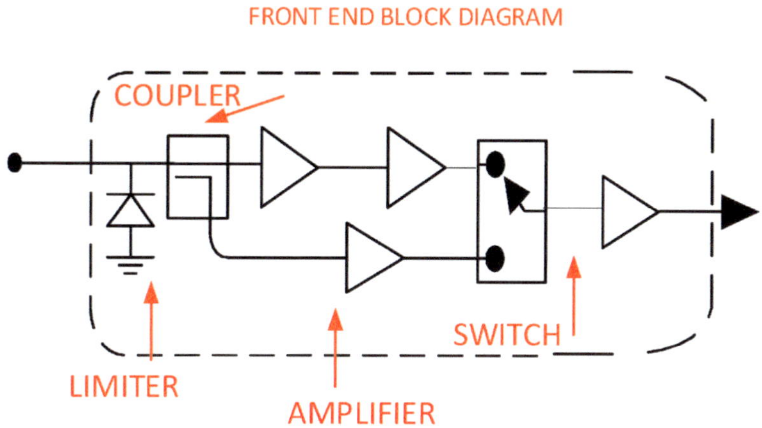

Figure 3.
MM-wave frontend block diagram.

Interface	Type
RF input	Wave guide WRD180 (double ridge)
RF output	K connector
DC supply	D type
Control	D type

Table 2.
Interface connectors.

protection, the die is passivated at the active areas. One series and two shunt PHEMTs per throw provide 2.0 dB insertion loss and 28 dB isolation at 40 GHz. The isolation between the SPDT input port to the output ports is better than 25 dB. The SPDT 1 dBc compression point is around 14 dBm. The SPDT dimensions are 1 x 0.7 x 0.1 mm. The frontend electrical characteristics was evaluated using ADS Keysight software [15] and SYSCAL software. The MMIC amplifiers and the SPDT are connected to the surface of the mechanical box. The MMIC chips are assembled on a CoVar carrier. During development it was found that the spacing between the frontend carriers should be less than 0.03 mm to achieve flatness requirements and V.S.W.R better than 2:1.

Figure 4 presents the frontend block diagram used to calculate the module noise figure and gain. The LNA noise figure is 6 dB. The frontend calculated noise figure is 9.46 dB. The frontend gain is 21 dB. The computed frontend noise figure and gain for LNA noise figure of 5.5 dB is presented in **Figure 5**. The module computed noise figure is 9.25 dB, with 21 dB gain.

The frontend current and voltage consumption are given in **Table 3**. The frontend may operate in high gain state or in low gain state. The gain of the high gain channel is higher by 15 to 20 dB than the low gain channel. The frontend measured gain is presented in **Figure 6**. The measured gain is around 20 ± 4 dB at the frequency range of 18 to 40 GHz.

2.4 High gain wideband frontend module

To achieve a high gain frontend module a medium power MMIC amplifier, HMC283, was added to the frontend module as presented in **Figure 7**. The HMC283 chip is a medium power amplifier. The chip consists of a four-stage GaAs amplifier

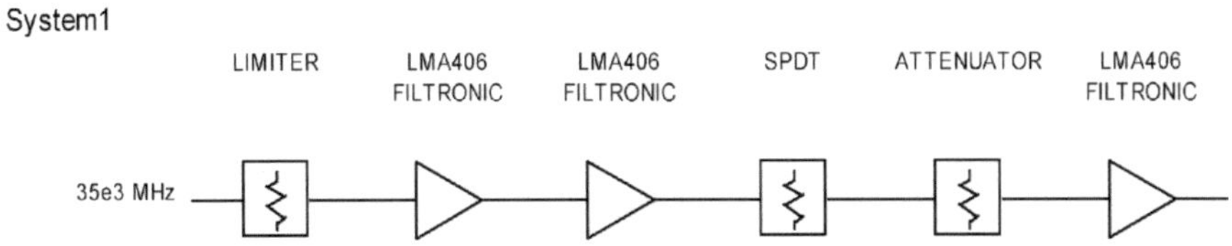

							Total
NF (dB)	3.00	6.00	6.00	3.00	3.00	6.00	9.46
Gain (dB)	-3.00	10.00	10.00	-3.00	-3.00	10.00	21.00
OIP3 (dBm)	30.00	25.00	25.00	30.00	30.00	25.00	23.14

Input Pwr (dBm)	-60.00	System Temp (K)	290.00	IM Offset (MHz)	.025
OIP2 (dBm)	23.14	OIP3 (dBm)	23.14	Output P1dB (dBm)	11.37
IIP2 (dBm)	2.14	IIP3 (dBm)	2.14	Input P1dB (dBm)	-8.63
OIM2 (dBm)	-101.14	OIM3 (dBm)	-163.28	Compressed (dB)	0.00
ORR2 (dB)	62.14	ORR3 (dB)	124.28	Gain, Actual (dB)	21.00
IRR2 (dB)	31.07	IRR3 (dB)	41.43		
SFDR2 (dB)	61.34	SFDR3 (dB)	81.79	Gain, Linear (dB)	21.00
AGC Controlled Range:		Min Input (dBm)	N/A	Max Input (dBm)	N/A

Figure 4.
Frontend module design for LNA NF = 6 dB.

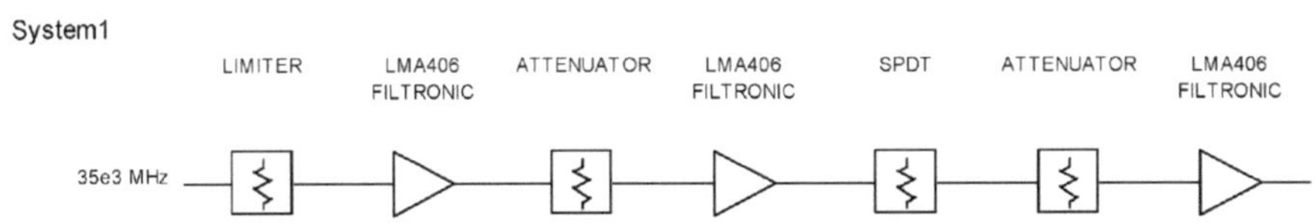

								Total
NF (dB)	3.00	5.50	3.00	5.50	3.00	1.00	5.50	9.25
Gain (dB)	-3.00	10.50	-3.00	10.50	-3.00	-1.00	10.00	21.00
OIP3 (dBm)	30.00	25.00	30.00	25.00	30.00	30.00	25.00	23.61

Input Pwr (dBm)	-60.00	System Temp (K)	290.00	IM Offset (MHz)	.025
OIP2 (dBm)	24.04	OIP3 (dBm)	23.61	Output P1dB (dBm)	11.74
IIP2 (dBm)	3.04	IIP3 (dBm)	2.61	Input P1dB (dBm)	-8.26
OIM2 (dBm)	-102.04	OIM3 (dBm)	-164.23	Compressed (dB)	0.00
ORR2 (dB)	63.04	ORR3 (dB)	125.23	Gain, Actual (dB)	21.00
IRR2 (dB)	31.52	IRR3 (dB)	41.74		
SFDR2 (dB)	61.90	SFDR3 (dB)	82.24	Gain, Linear (dB)	21.00
AGC Controlled Range:		Min Input (dBm)	N/A	Max Input (dBm)	N/A

Figure 5.
Frontend module design for LNA NF = 5.5 dB.

Voltage (v)	3	5	−12	−5	5 Digital
Current (A)	0.25	0.15	0.1	0.1	0.1

Table 3.
Wideband frontend module voltage and current consumption.

MMIC in the frequency range of 17 to 40 GHz. The chip dimension is 1.62 mm x 1.62 mm and may be integrated into multi-chip modules. The chip is a GaAs PHEMT amplifier with around 20 dB gain and + 21 dBm output power. The amplifier bias voltage is +3.5 V and consumes 300 mA of current.

The amplifier gain is around 21 dB with 10 dB noise figure and 21 dBm saturated output power. The amplifier dimensions are 1.72 x 0.9 mm. The high gain frontend module block diagram is shown in **Figures** 7 and **8**. The frontend module consists of high gain and low gain channels. **Figure** 7 presents the high gain and low gain channels without an amplifier in the low gain channel. **Figure 8** presents the high

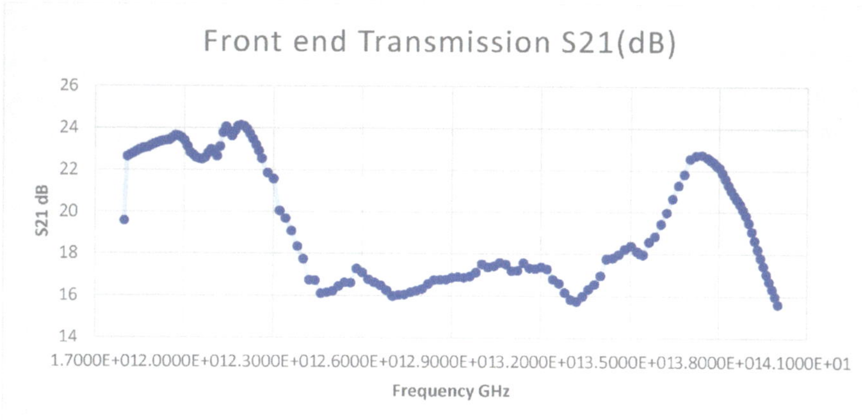

Figure 6.
Measured wideband frontend gain 17 GHz to 40 GHz.

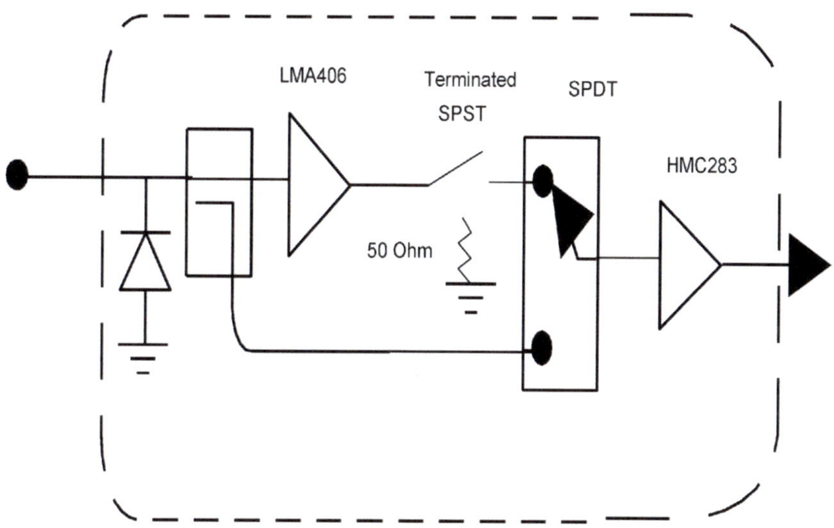

Figure 7.
High gain frontend block diagram.

gain and low gain channels with an amplifier in the low gain channel. The gain difference between the high gain and low gain channels for the frontend presented in **Figure 8** is around 10 to 15 dB. Block diagram of the wideband frontend with one LMA406 amplifier in the high gain channel is shown in **Figure 9**.

2.5 High gain frontend design

The high frontend electrical characteristics were evaluated using ADS Agilent software and SYSCAL software. The calculated noise figure and gain for LNA noise figure of 9.5 dB is presented in **Figure 10**. The frontend calculated noise figure is 13.3 dB. The frontend gain is 32.5 dB. The frontend module calculated noise figure and gain for LNA noise figure of 5 dB is presented in **Figure 11**. The frontend calculated noise figure is 10 dB. The frontend gain is 29.5 dB.

Table 4 lists the measured results of frontend modules. **Figure 12** is a photo of HMC283 amplifier assembly, and **Figure 13** is a photo of the wideband compact frontend. **Table 5** provides the performance and cost of commercial mm-wave

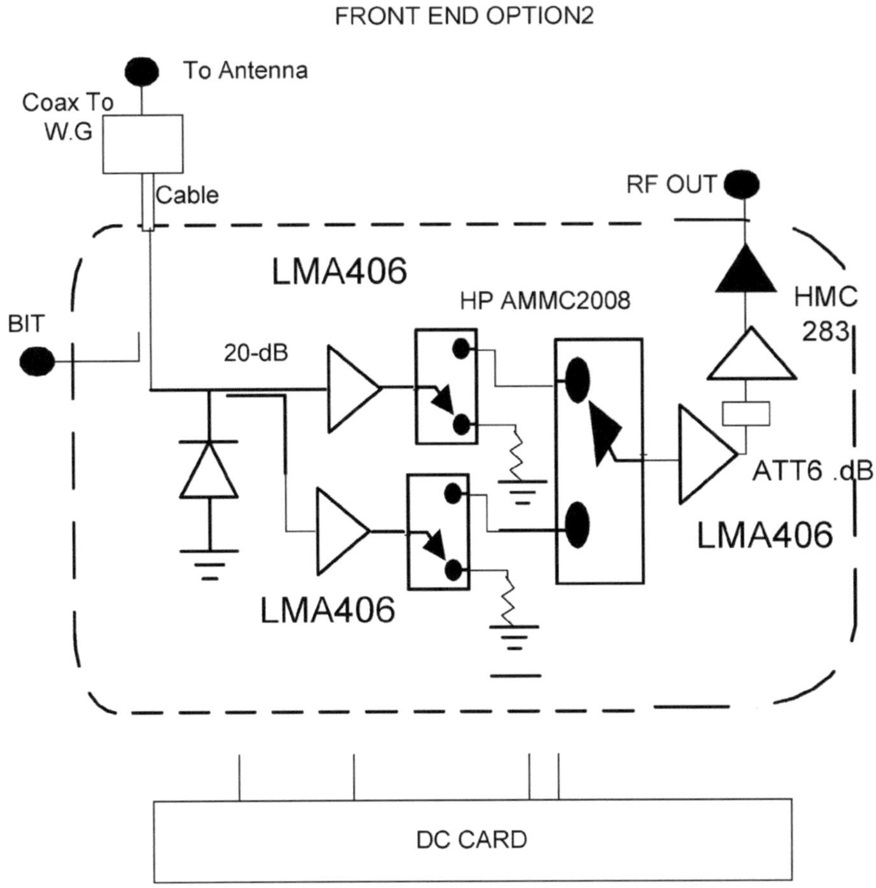

Figure 8.
High gain frontend block diagram with amplifier in the low gain channel.

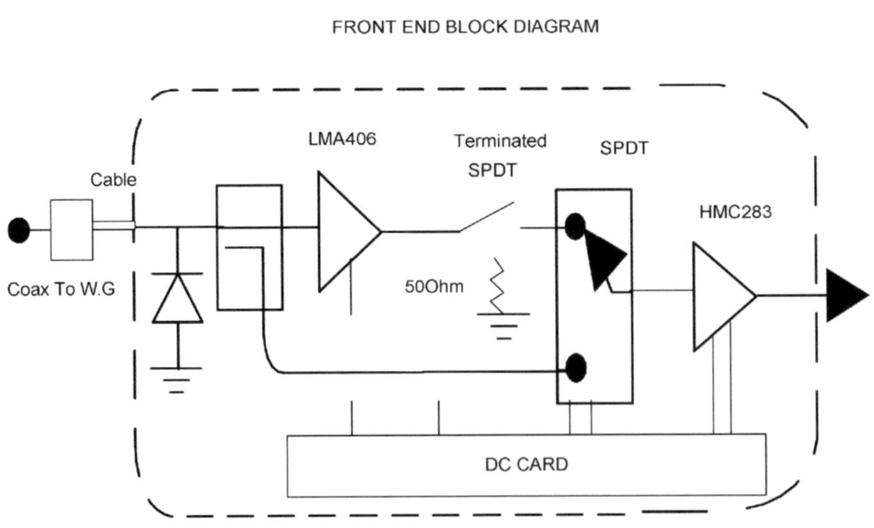

Figure 9.
Block diagram of the high gain wideband frontend with one LMA406 amplifier.

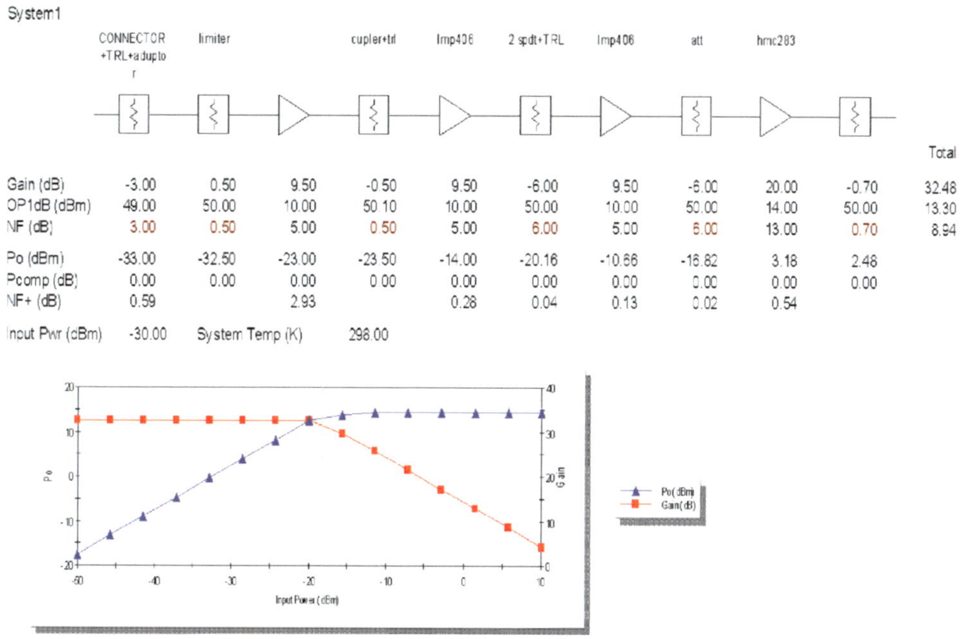

Figure 10.
Wideband frontend module design for LNA NF =9.5 dB.

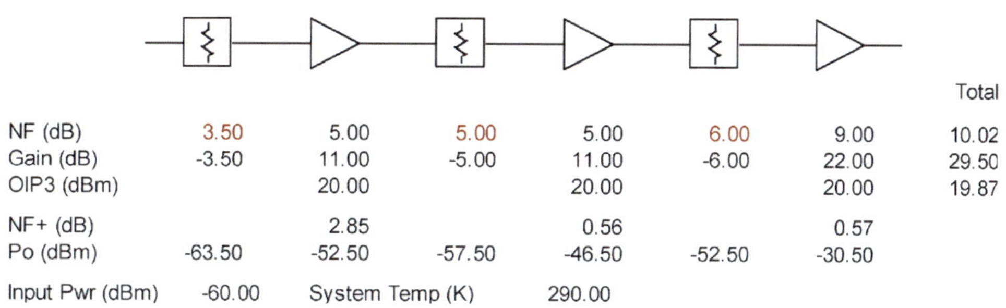

Figure 11.
Wideband frontend module design for LNA NF = 5 dB.

Parameter	OMNI02	OMNI01	DF6	DF4	DF3	DF2	DF1
Max. high gain mode	31.5	32	31	32	32.5	32.5	31
Min. high gain mode	28.5	28	26	26	27.5	27	26
Avg. high gain mode	30	30	29	29	29	29	29
Amplitude balance	3	4	5	6	5	5	4
S11 (dB)	5	4	4.5	5	5	5	5
S22 (dB)	7	6	7.5	6	5	6	5
Isolation (dB)	21.5	22.5	9	9	10	10	6.5
Max. low gain mode	16.5	18	19	18	17	17	17
Min. low gain mode	10.5	11	13	10	7.5	12	12
Avg. low gain mode	13.5	14.5	15	14	12	14	14
Amplitude balance	6	7	6	8	9.5	5	5

Parameter	OMNI02	OMNI01	DF6	DF4	DF3	DF2	DF1
P1 dBC 30 GHz	14	15.96	11.6	11.93	11.7	11.4	10.9
P1 dBC 40 GHz	14.48	16.8	13.96	14.5	15.58	15.28	14
NF 30 GHz	8.14	8.75	8.68	9.48	8.65	8.45	10.5
NF 40 GHz		8.75	9.28	10.1	8.64	9.17	10.24

Table 4.
Measured results of the wideband frontend modules.

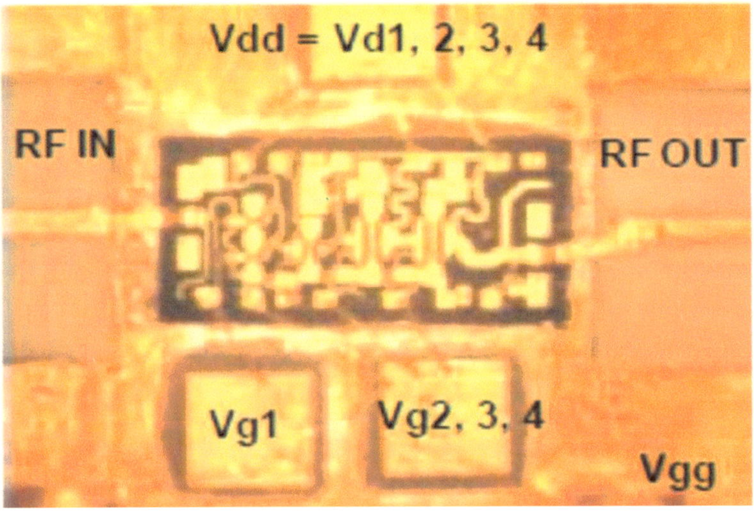

Figure 12.
Photo of HMC283 amplifier assembly.

Figure 13.
Compact 18–40 GHz wideband frontend module.

microwave components. **Table 6** lists the costs of commercial MMIC and MIC mm-wave microwave components. There is a good agreement between computed and measured results. **Figure 14** is a photo of the compact mm-wave wideband, 18–40 GHz, frontend.

Component	Model#	Freq. [GHz]	NF / IL [dB]	Gain dB]	Power out 1dBc [dBm]	Coupling [dB]	Isolation [dB]	VSWR max	Unit Cost [K$]
Switch General Microwave	F9022	18–40	4	N/A		N/A	40	2.30:1	2.1
Coupler KRYTAR	184020	18–40	1.3	N/A		20		1.70:1	0.8
LNA Space Labs	SLKKa-30-6	18–40	6	30	17	N/A		1.80:1	2.2
WG/K Adapter MICROTEHC	R45 240130	18–40 18–40	0.75					1.25:1 1.6:1	0.5

Table 5.
Performance and cost of available connectorized mm-wave RF components.

Application	Component	Model#	TYPE	Unit Cost $
RF Frontend	Switch General Microwave	F9022	Drop in	1800
RF Frontend	LNA Filtronic	LMA406	MMIC	24
RF Frontend	LNA/VGA UMS	CHA2097A	MMIC	42
RF Frontend	Limiter Diode	GC4800	Diode	100
SFB	Switch	HMMC2027 HP	MMIC	80
SFB	Switch	TLCSWY04	MMIC	80

Table 6.
Cost of available mm-wave RF components.

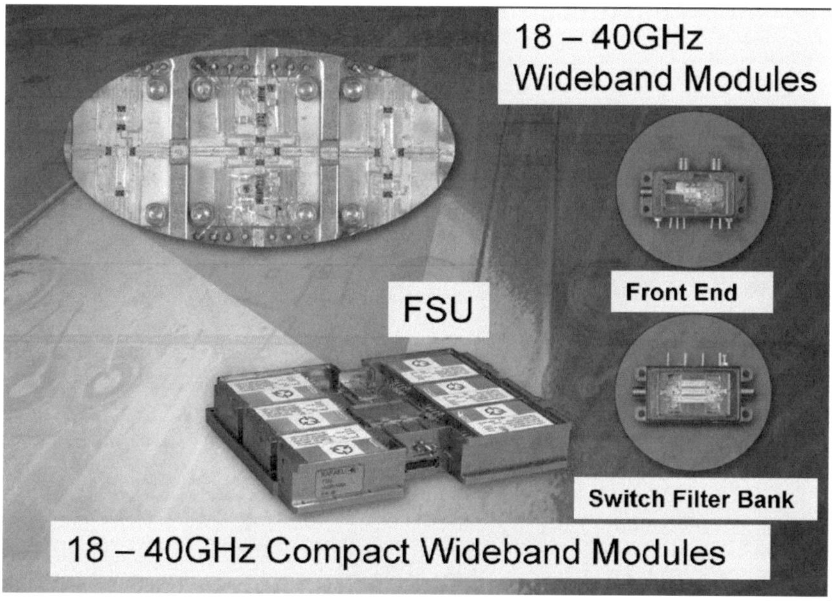

Figure 14.
Photo of 18–40GHz compact modules.

3. Wideband, 18 to 40 GHz, integrated compact switched filter Bank module

3.1 Introduction

Filter bank modules are an important module in direction finding, radar, seeker, and communication systems. The electrical performance of the filter bank determines if the system will meet the system-required dynamic range and signal-to-noise ratio (SNR) specifications. Moreover, in several cases the filter bank performance limits the system dynamic range. This section describes the design and development of an integrated, low-cost, 18 to 40 GHz wideband compact filter bank module. Design and fabrication considerations are presented.

The wideband SFB consists of three wideband side-coupled microstrip filters. The SFB MIMIC switches operate in the 18 to 40 GHz frequency range and are used to select the required filter. The insertion loss of each filter section is less than 11.5 dB $\pm$ 1.5 dB. The passband bandwidth of each filter is around 8 GHz. The received signal is rejected by 40 dB at $\pm$7GHz from the center frequency. The received signal is rejected by 60 dB at $\pm$11GHz from the center frequency. The SFB volume is 2 x 5 x 1 cm.

3.2 Description of the filter bank

Figure 15 presents the block diagram of the compact SFB unit. The SFB The module consists of three side-coupled microstrip filters. Each side-coupled filter consists of nine sections. The filters are printed on a 5-mil alumina substrate. One to two dB attenuators connect the filters input and output ports to wideband MMIC SPDT switches. The attenuators are used to adjust each channel's losses to the average required level. The module losses are adjusted to be higher in the low frequencies and lower in the high frequencies. The adjustment of the attenuation level improves the filters flatness over a wideband, 18–40GHz frequency range. The filters are assembled to the surface of the package metal box. The SFB switches are assembled on a CoVar carrier. In the development process of the SFB unit, we found

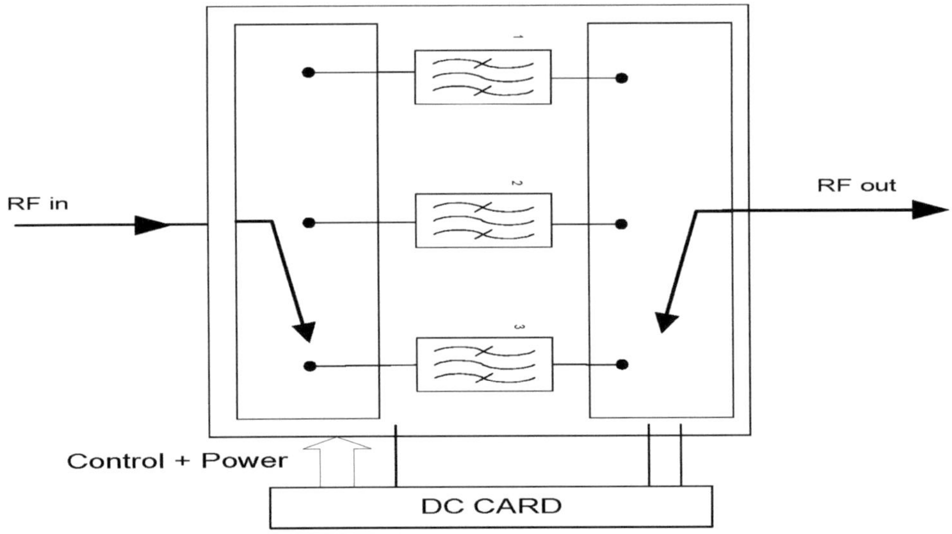

Figure 15.
Block diagram of the compact filter bank module.

that the spacing between the filter carriers and the switch carriers should be less than 0.03 mm to get S11 and S22 better than −9.5 dB and flatness better than ±1 dB.

3.3 Wideband switch filter Bank specifications

Table 7 lists the specifications of the wideband SFB.

3.4 Wideband filter design

The filters were designed by employing AWR and ADS software. **Figure 16** presents single filter response requirements. **Table 8** and **Figure 17** show the SFB expected frequency response. **Table 9** presents the advantages of the integrated design over the discrete design. For example, the weight of a discrete SFB is 1 kg and the weight of an integrated SFB is 50 g. The volume of the discrete SFB is twice the volume of the integrated SFB. The filter contains nine side-coupled microstrip lines printed on a 5-mil alumina substrate. ADS and AWR software were applied to optimize the filter dimensions and structure to meet the system requirements. **Figures 18–20** show the computed results of the filters. The sensitivity of the design to substrate tolerances such as height and dielectric constant has been optimized by using RF analysis software, see (**Figure 21**). We fabricated the filter configuration that was less sensitive to production tolerances (**Figure 24**).

Figure 22 presents computed S_{11} and S_{21} parameters for the SFB using ADS software. **Figure 23** presents the expanded S12 computed results of the filter bank. **Figure 24** presents the SFB board drawing. **Figure 25** is a photo of the SFB. **Figure 26** presents the S21 SFB results of the first unit measured during the production process. A comparison of the SFB computed and measured results proves that there is a good agreement between computed and measured results.

Parameter	Requirements
Frequency range	18–40.1 GHz
Number of channels	3
Channel Passband	8 GHz
Rejection	40 dB min @ F0 ± 8GHz 50 dB min @ F0 ± 11GHz
Flatness	± 1.2 dB max. For 4 GHz
I.L CH-1 CH-2 CH-3	12–14.5 dB 10.5–13.5 dB 9–12 dB
Switching time	100 nsec
Input Power	25 dBm max
VSWR	2.5:1 max
Control	2 LVTTL lines
Power Supply voltages	± 5 V DC. Heat dissipation – 1 W max
Dimension	5 x 2 x 1 cm

Table 7.
Wideband SFB specifications.

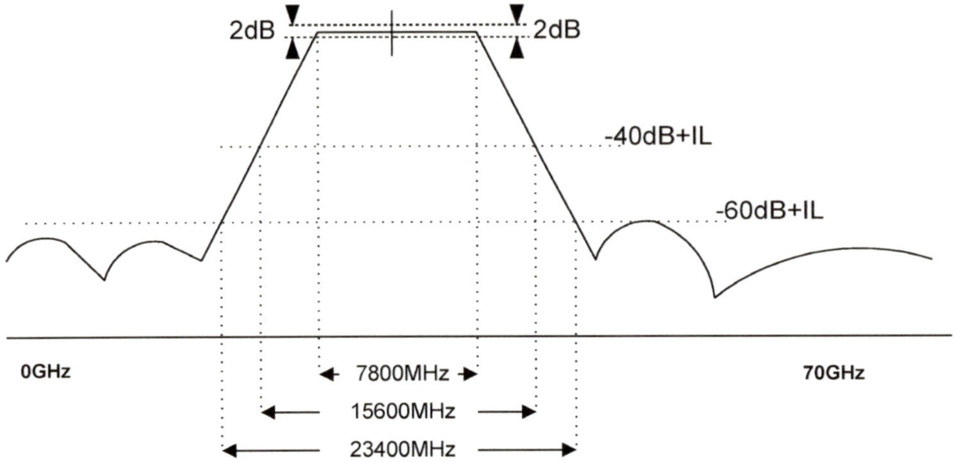

Figure 16.
Single filter response requirements.

CH	Rejection −60 dB	Rejection −40 dB	Passband −3 dB	Passband −3 dB	Rejection −40 dB	Rejection −60 dB
CH-1 (GHz)	10.1	14	17.9	25.7	29.6	33.5
CH-2 (GHz)	17.3	21.2	25.1	32.9	36.8	40.7
CH-3 (GHz)	24.5	28.4	32.3	40.1	44	47.9

Table 8.
SFB requirements.

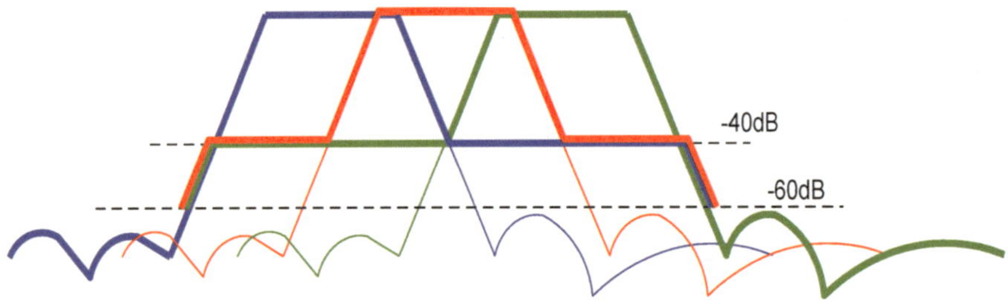

Figure 17.
SFB expected frequency response.

Parameter design	Dimension Cm	Weight Kg.	Price K$
Integrated	5.5 x 2.5 x 1.5	0.05	2.2
Discrete	12 x 6 x 3	1	10

Table 9.
Comparison between discrete and integrated design.

Figure 27 shows the measured S12 parameters of filter #2 during the production process. **Figure 28** shows the measured S12 parameters of the SFB during the production process. The SFB losses at high frequencies are around 9 dB and at the low frequencies the losses are around 9 dB (**Figure 28**). **Figure 29** presents the detailed measured S11 parameter of the SFB. **Figure 30** presents the detailed measured S12 parameter of the SFB.

CH-1 FILTER

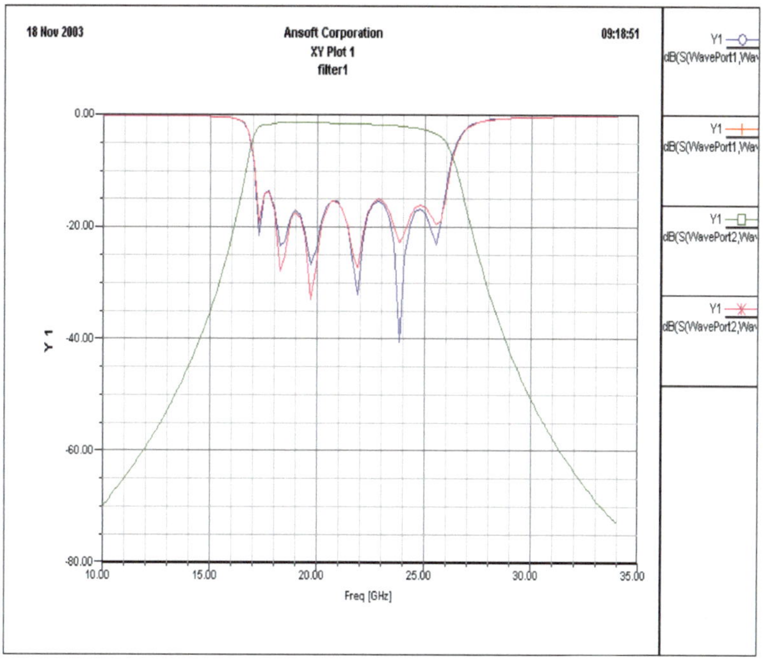

Figure 18.
S parameters for filter #1.

CH-2 FILTER

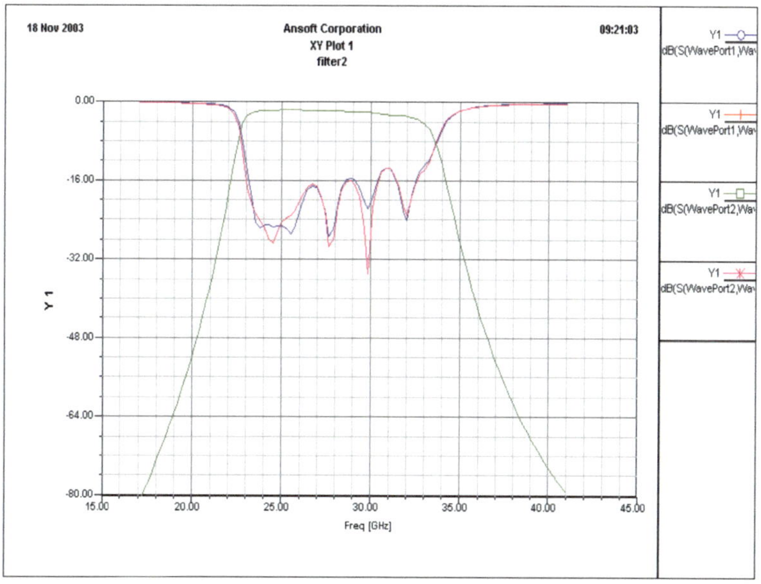

Figure 19.
S parameters for filter #2.

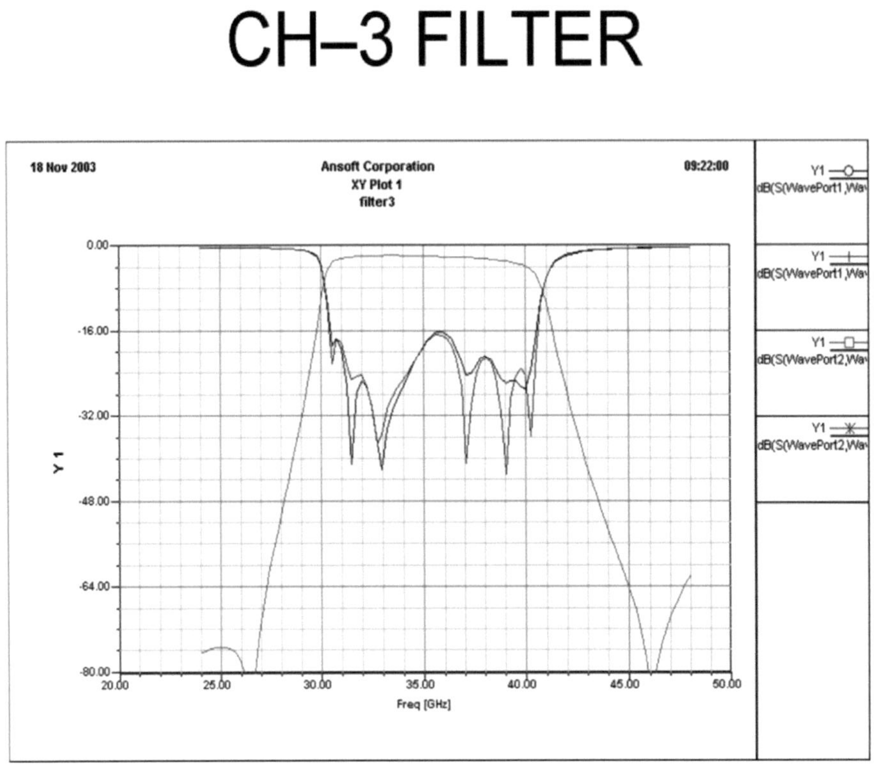

Figure 20.
S parameters for filter #3.

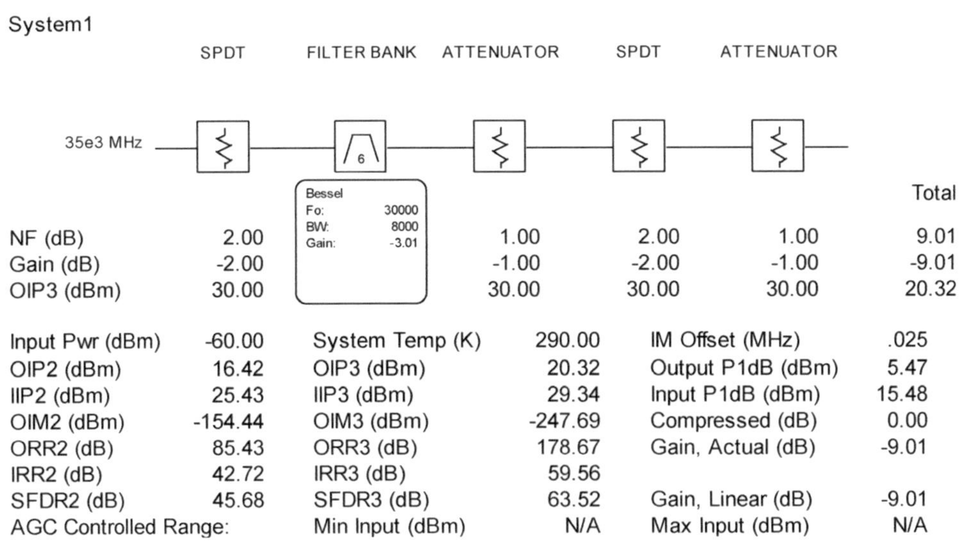

Figure 21.
SFB analysis results.

4. Wideband MM-Wave Couplers

Couplers are used to couple part of the power in the input port to a coupled port. For example, as presented in **Figures 8** and **9**. Usually couplers consist of two coupled quarter-wavelength transmission lines and have four ports, as shown in

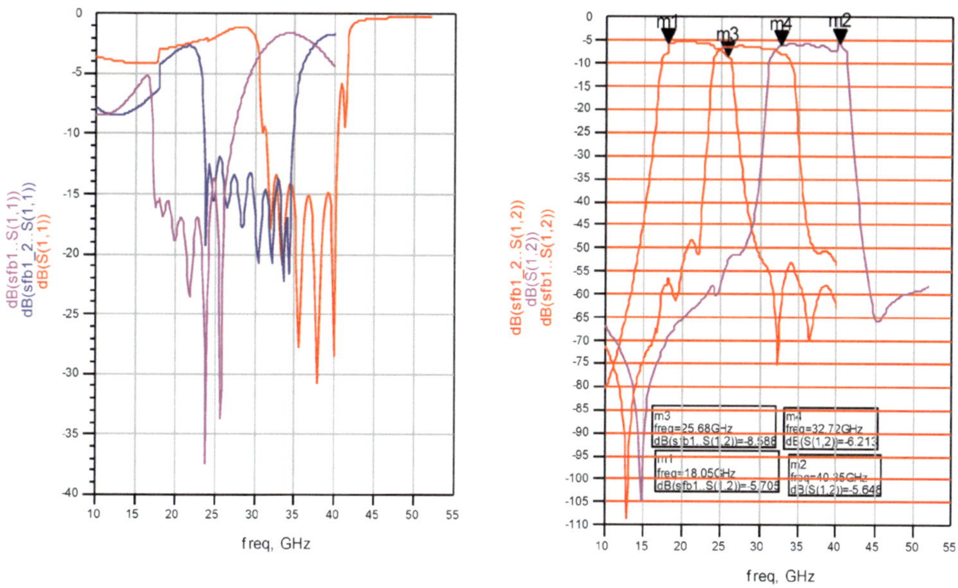

a

b

Figure 22.
a. S_{11} SBF computed results; b. S_{21} SBF computed results.

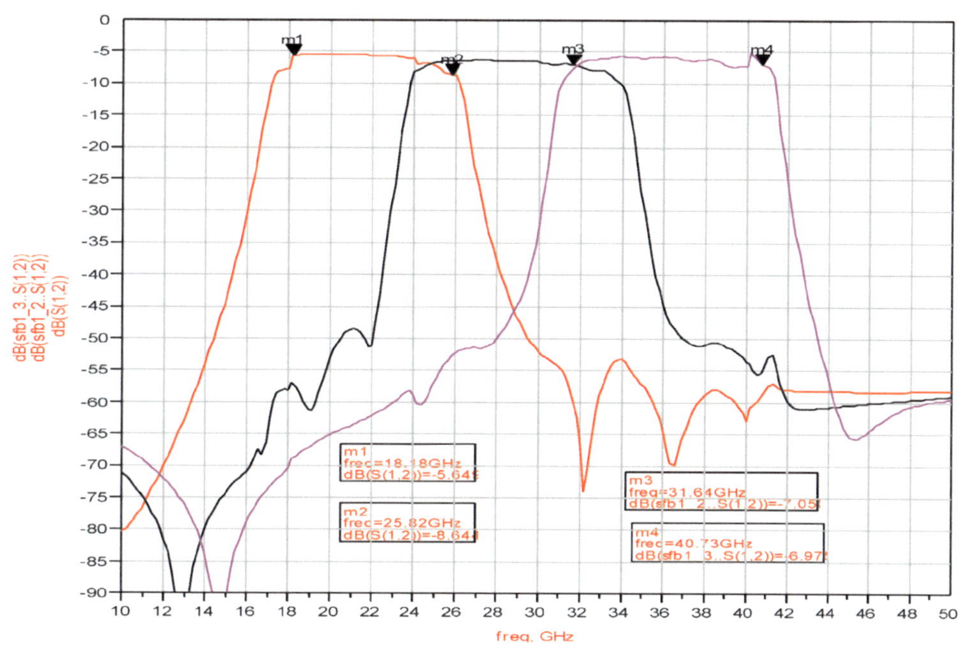

Figure 23.
SFB S12 computed results.

Figure 31. P1 is the input port, P2 is the transmitted port, P3 is the coupled port, and P4 is the isolated port. The coupled port may be used to obtain the information about the signal such as frequency and power level without interrupting the main power flow in the device. The coupling factor may be calculated by using Eq. (1) and is the ratio between the coupled power to the input power in dB when the other ports are terminated. Coupler losses may be calculated using Eq. (2). The overall

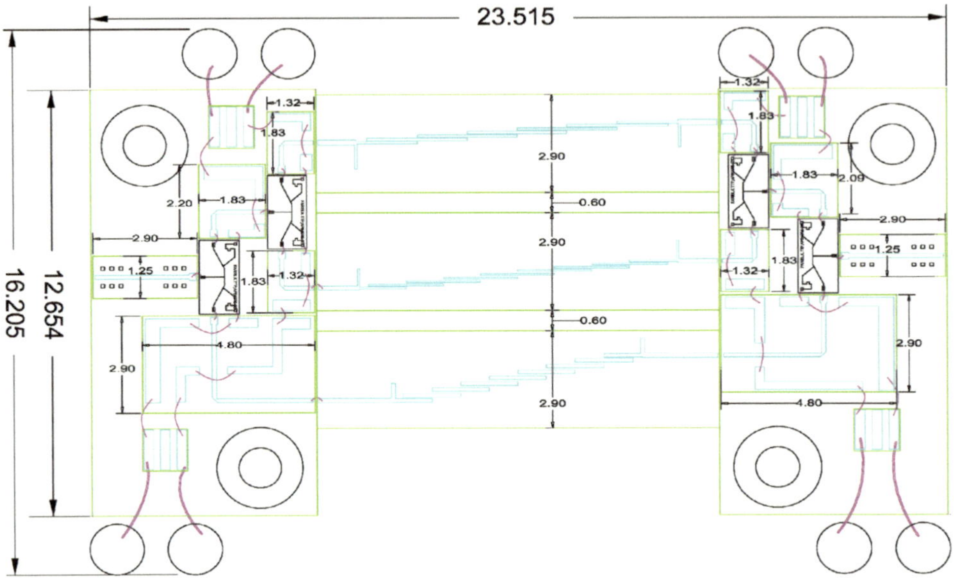

Figure 24.
SFB layout.

Figure 25.
SFB picture.

coupler losses are due to conductor losses, coupling losses, dielectric losses, radiation losses, and matching losses.

$$Coupling\ Factor = CF = -10\,log\,\frac{P3}{P1} \tag{1}$$

$$Insertion\ Loss = IL = 10\log\left(1 - \frac{P3}{P1}\right) \tag{2}$$

The isolation factor is the ratio between the power in the isolated port to the input power in dB when the other ports are terminated and is given by Eq. (3).

$$Isolation = -10\,log\,\frac{P4}{P1} \tag{3}$$

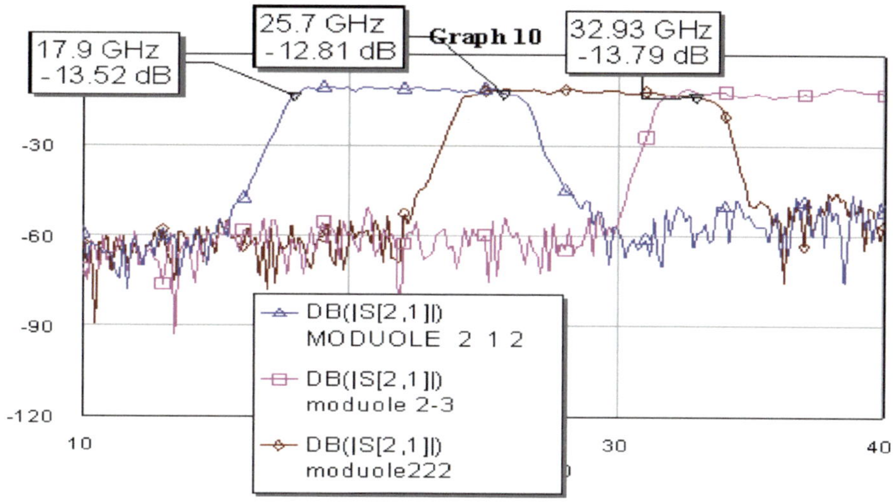

Figure 26.
SFB measured results unit number 1.

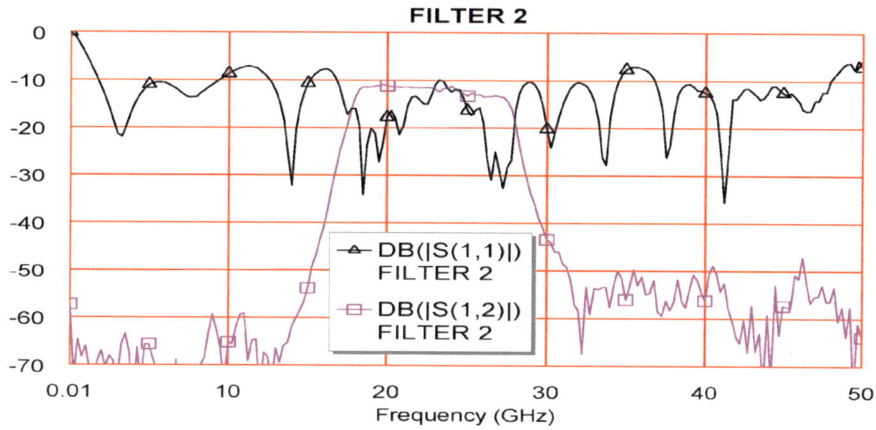

Figure 27.
Measured S parameters for filter #2.

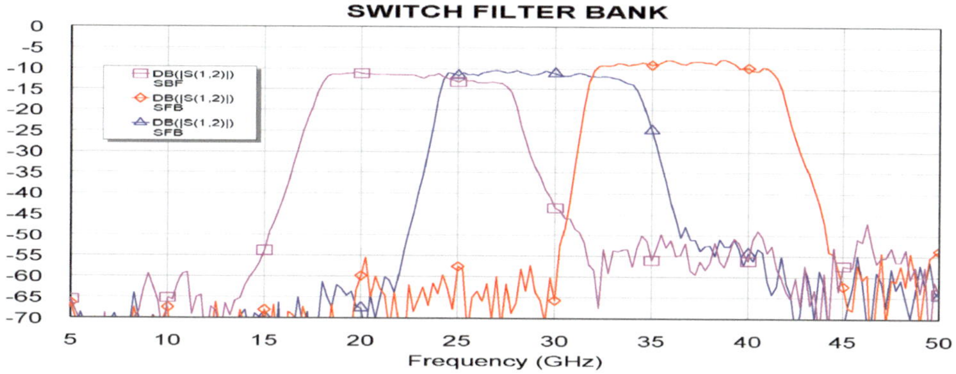

Figure 28.
SFB measured S12 results.

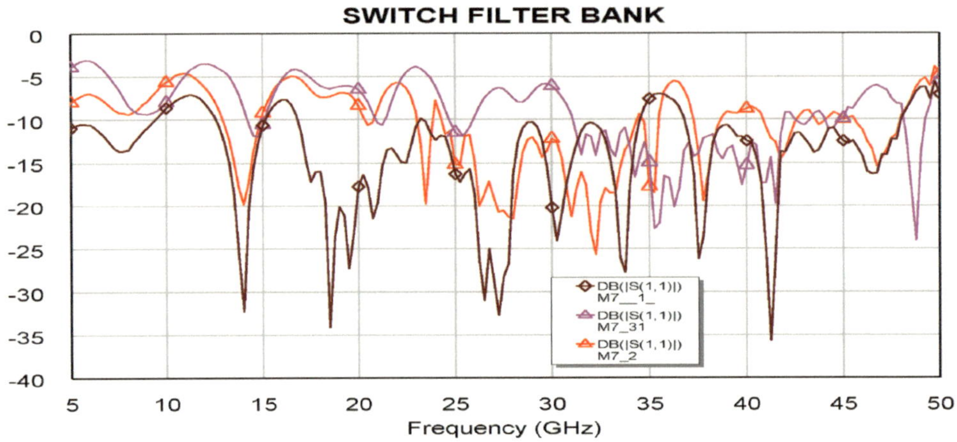

Figure 29.
SFB measured S11 results.

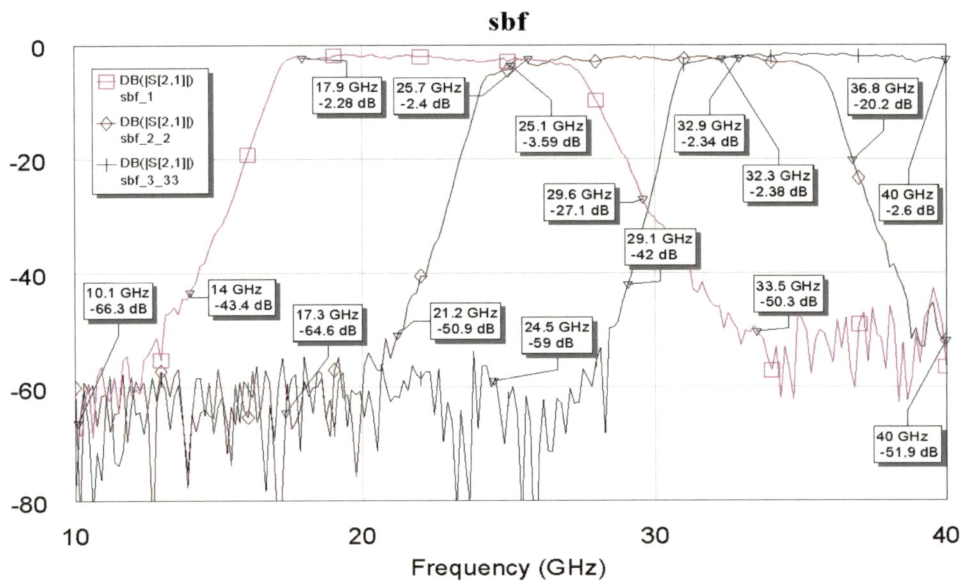

Figure 30.
SFG detailed measured S12 results.

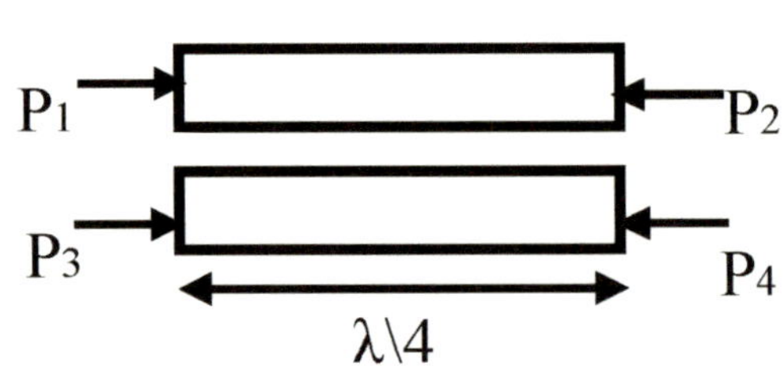

Figure 31.
Coupled lines.

DOI: http://dx.doi.org/10.5772/intechopen.97853

The coupler directivity is the ratio between the power in the isolated port to the coupled power in dB when the other ports are terminated and is given by Eq. (4).

$$Directivity = D = -10\,log\,\frac{P4}{P3} \tag{4}$$

In radars, DF systems, seekers, and communication systems, the amplitude and phase balance between the system ports determines the accuracy of the signal processing process. The power difference in dB between two output ports of the system is defined as the amplitude balance. In an ideal hybrid circuit, the amplitude difference between two ports should be 0 dB. The phase difference in degrees between two output ports of the system is defined as the phase balance. However, in real life the amplitude balance and phase balance vary with frequency.

4.1 A wideband MM-wave coupler

Figure 32 presents a wideband MM-wave coupler printed on alumina with a 9.8 dielectric constant and 5.5 mil-inch thickness. The coupler frequency range is 18 GHz to 40 GHz. The coupler was designed by Momentum ADS software. The coupling value is −13 dB and is shown in **Figure 33**. The coupler insertion loss is 0.4 dB and is shown in **Figure 34**. The coupler S11 parameter is better than −26 dB and is shown in **Figure 35**.

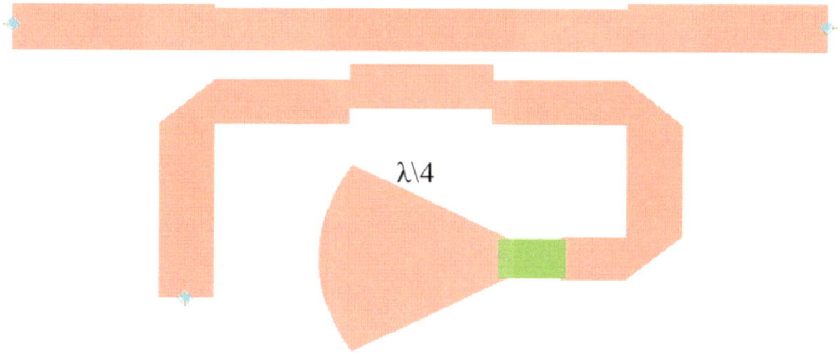

Figure 32.
18 GHz to 40 GHz, wideband MM-wave coupler.

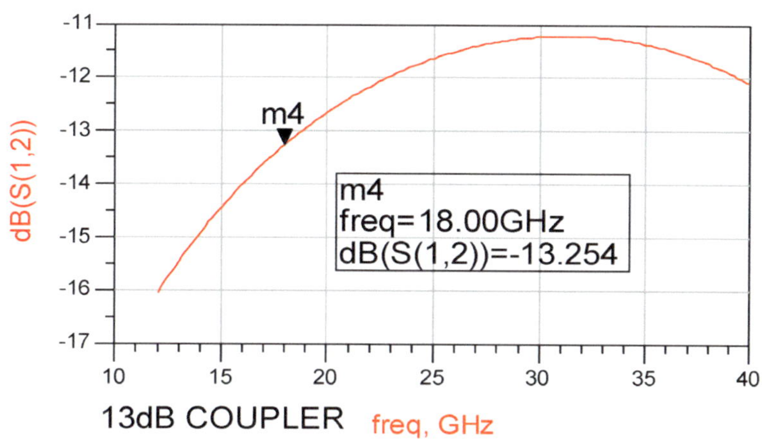

Figure 33.
The wideband coupler coupling value, S12.

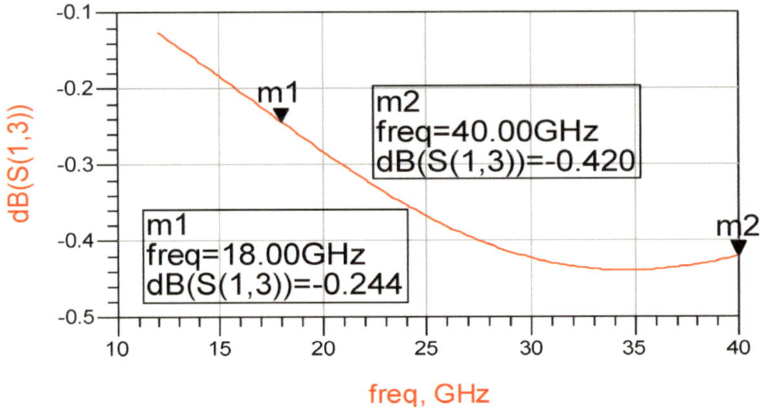

Figure 34.
The wideband coupler insertion loss.

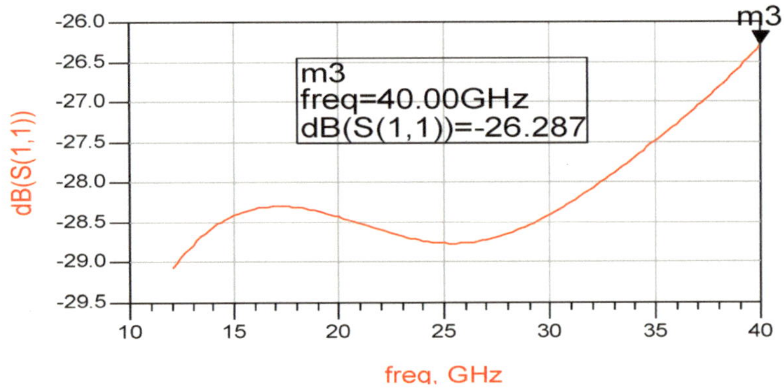

Figure 35.
S11 of the wideband coupler.

5. Conclusions

This chapter presented a low-cost, integrated 18–40 GHz wideband compact frontend with a 47 dBm high power limiter. The frontend consists of two channels: a high gain and low gain channel. Wideband MMIC switches are employed to select the required channel. The gain of the high gain channel is around 27 dB with ±1 dB flatness. The noise figure of the module is around 9 dB.

This chapter also discussed the design and performance of a compact, low-cost, integrated SFB. The integrated SFB has several advantages over the discrete SFB. For example, the weight of a discrete SFB is 1 kg and the weight of an integrated SFB is 50 g. The volume of the discrete SFB is twice the volume of the integrated SFB. The filter contains nine side-coupled microstrip lines printed on a 5-mil alumina substrate. ADS and AWR software were applied to optimize the filter dimensions and structure to meet the system requirements. The filter's passband bandwidth is around 8 GHz. The SFB insertion loss is around 12 dB in the low frequency range and 9.5 dB in the high frequency range. The filter flatness in the passband frequencies is better than ±1 dB. The received signal is rejected by 40 dB at ±7 GHz from the center frequency. The received signal is rejected by 60 dB at ±11GHz from the center frequency. The SFB volume is 2 x 5 x 1 cm.

Author details

Albert Sabban
Kinneret College, Kinneret Israel and Ort Braude College, Karmiel, Israel

*Address all correspondence to: sabban@netvision.net.il

References

[1] J. Rogers, C. Plett "Radio frequency Integrated Circuit Design", Artech House, 2003.

[2] N. Maluf, K. Williams, "An Introduction to Microelectromechanical System Engineering", Artech House, 2004.

[3] Albert Sabban (2016). "Wideband RF Technologies and Antenna in Microwave Frequencies", Wiley Sons, July 2016, USA.

[4] Albert Sabban Editor and Author (2020). Wearable Systems and Antennas Technologies for 5G, IOT and Medical Systems CRC PRESS, TAYLOR & FRANCIS GROUP, December 2020. ISBN 9780367409135.

[5] Albert Sabban (2017). "Novel Wearable Antennas for Communication and Medical Systems", TAYLOR & FRANCIS GROUP, October 2017.

[6] Albert Sabban (2015). "Low visibility Antennas for communication systems", TAYLOR & FRANCIS GROUP, 2015, USA.

[7] S. A. Mass, "Nonlinear Microwave and RF Circuits", Artech House, 1997.

[8] Abdeen Hebat-Allah Yehia; Yuan Shuai; Schumacher Hermann; Ziegler Volker; Meusling Askold; Feldle Peter. "10 to 40 GHz Superheterodyne Receiver Frontend in 0.13 µm SiGe BiCMOS Technology". Frequenz, Volume 71, pp.151-160. March 2017.

[9] A. Sabban, "Microstrip Antenna Arrays", 2011, Microstrip Antennas, Nasimuddin Nasimuddin (Ed.), ISBN: 978-953-307-247-0, InTech, pp. 361-384,

[10] A. Sabban, "Applications of MM Wave Microstrip Antenna Arrays" ISSSE 2007 Conference, Montreal Canada, August 2007.

[11] Reuven Shavit, Albert Sabban, Michael Sigalov, Avihai Lahman, Zeev Iluz, Naftali Chayat and Solon Spiegel "microwave engineering research activity in Israel", IEEE Microwave Magazine, May 2018, 19(3):129-135. DOI: 10.1109/MMM.2018.2802281

[12] G.P. Gauthier, G.P. Raskin, G.M. Rebiez., P.B. Kathei, "A 94GHz Micro-machined Aperture- Coupled Microstrip Antenna", I.E.E.E Trans. on Antenna and Propagation Vol. 47, No. 12, pp. 1761-1766, December 1999.

[13] G. de Lange et. al., "A 3*3 mm-wave micro-machined imaging array with sis mixers", Appl. Phys. Lett. 75 (6), pp. 868-870, 1999.

[14] A. Rahman et. al., "Micro-machined room temperature micro bolometers for MM-wave detection", Appl. Phys. Lett. 68 (14), pp. 2020-2022, 1996.

[15] A. Sabban and K.C. Gupta, "Characterization of Radiation Loss from Microstrip Discontinuities Using a Multiport Network Modeling Approach", I.E.E.E Trans. on M.T.T, Vol. 39, No. 4, April 1991, pp. 705-712.

[16] Keysight software, http://www.keysight.com/en/pc-1297113/advanced-design-system-ads?cc=IL&lc=eng

Chapter 3

UWB-MIMO Antenna with Band-Notched Characteristics for Portable Wireless Systems

Chandrasekhar Rao Jetti and Venkateswara Rao Nandanavanam

Abstract

The current and future wireless communication systems demand for higher data rates, enhanced quality of service and more channel capacity. Since Federal Communications Commission (FCC) allocated the unlicensed frequency spectrum from 3.1 to 10.6 GHz for commercial applications in the year 2002, Ultra-Wideband (UWB) technology has attained considerable attention because of its inherent features like high data rate, more channel capacity, extremely less power consumption and low cost. However, for UWB systems, multipath fading and frequency interference are the two significant issues that requires further investigation. In recent times, Multiple Input Multiple Output (MIMO) technology has gained much attention in wireless communication as it exploits multipath to increase the communication range and link quality. Thus, MIMO technology is a promising solution for mitigating multipath fading in UWB system. However, accommodating multiple antennas with less mutual coupling between them in portable devices is always a challenging task for antenna designers. UWB system could easily interfere with existing narrowband communication system such as Wireless Local Area Network (WLAN). So, the design of an ultra-wideband antenna with integrated frequency notching function is a good solution to suppress the frequency interference and to reduce the complexity of the UWB system instead of using a conventional filter. In this chapter, compact isolation-enhanced planar UWB-MIMO antenna with single band-notched characteristics is presented.

Keywords: ultra-wideband, multipath fading, frequency interference, multiple input multiple output, mutual coupling, isolation, band-notch

1. Introduction

The primary concerns of wireless communication systems are higher data rates, improved quality of service, more channel capacity, low power consumption and less interference with other systems. Ultra Wideband technology is the most promising technology to meet the requirements of wireless communication system because of its advantages, such as higher data rate transmission, low cost, high security, and less power requirement [1]. However, multipath fading and frequency interference with other communication systems are the important problems should be well solved for UWB systems.

In an indoor communication application, like other wireless communication systems, the UWB system performance is also restricted by multipath fading due to rich scattering environments which cause inter-symbol interference. In present times, digital communication using multiple input multiple output (MIMO) technology has emerged as a breakthrough for a wireless system. The MIMO system employs multiple antennas at the transmitter and receiver. It makes use of the rich multipath environment to mitigate the multipath fading effect. And it improves the range of communication and system capacity (data rate) without the need for additional bandwidth or transmitted signal power [2, 3]. Hence, the UWB system with MIMO technology is a viable solution to reduce the multipath fading effect and to improve the quality of service, the range of communication and system capacity [4].

The electromagnetic interaction between the radiating (antenna) elements in multiple antenna or MIMO system is known as mutual coupling. The closely spaced antennas, especially in portable devices, inevitably cause strong mutual coupling between antennas. The mutual coupling is undesirable which causes fluctuations in the input impedance of individual antenna element i.e., impedance mismatch which degrades the radiation efficiency, deviations in antenna radiation pattern due to the high correlation between antenna signals and decreases the channel capacity of MIMO system. Since the mutual coupling has a considerable impact on the MIMO system performance, the reduction of mutual coupling between antennas and enhancement of isolation between ports is imperative. However, placing multiple antennas in a space-limited portable wireless device is a big challenge for antenna designers [5]. Hence, designing compact UWB-MIMO antenna exhibiting band-notch function and less mutual coupling is very much needed. Various designs were proposed in the recent years to suppress the effects of mutual coupling in ultra-wideband MIMO antennas [6–15]. Methods include, placing radiating elements perpendicular to each other and adding two long protruding stubs to ground [6], use of tree-like structure on the ground plane [7], etching a T-shaped slot and a line slot on the ground [8], adding a Y-shaped slot on the T-shaped protruded ground plane [9], placing two shorts at 45 degrees between the microstrip lines and in the opposite direction [10], protruding ground structure [11], T-shaped metallic stub [12, 13], adopting wideband neutralization line [14] and using modified ground structure along with T-shaped slot on the ground [15].

Ultra-wideband is an emerging technology for short distance low power communications. It makes use of short duration pulses which have very low power spectral density for transmission of data. Since the UWB system is operating from 3.1 to 10.6 GHz, it could easily interfere with existing narrowband communication system like Wireless Local Area Network (WLAN-5.15–5.825 GHz). So, ultra-wideband antenna with integrated frequency notching function at the interfering frequency band is a feasible solution to mitigate the frequency interference [16]. The power spectral density of UWB and other narrowband systems is shown in **Figure 1**. The frequency interference produced by a UWB transmitter to a narrowband system is very negligible because of the transmitted signal emission power (power spectral density) is very less compared with narrowband systems. But, when a UWB receiver is located very near to the narrowband interferer, the interference caused is very high. So, a notch at the interfering frequency is needed to suppress its effect. The traditional RF filter circuits using lumped elements can be used to implement this frequency notching feature but, it increases the system complexity, cost and occupies more space when integrated with other microwave circuits in the portable device. Another viable solution is to design a UWB antenna with an integrated band-notched feature to mitigate the frequency interference which decreases the complexity and cost of the UWB system. The idea of designing

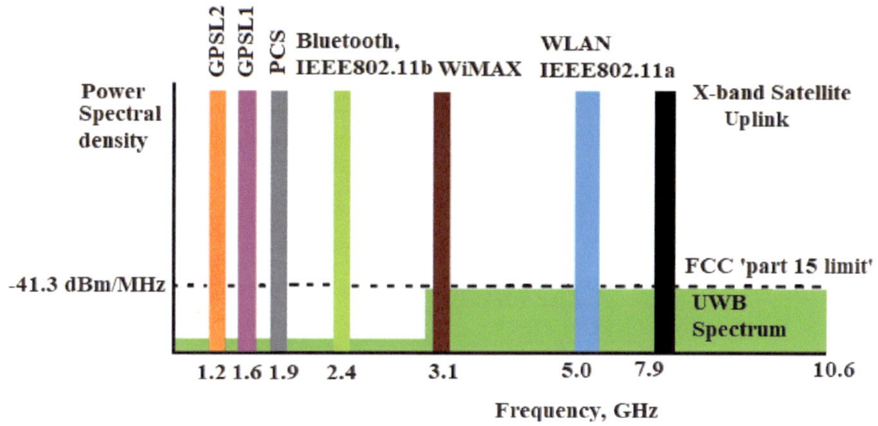

Figure 1.
The power spectral density of UWB and other narrowband systems.

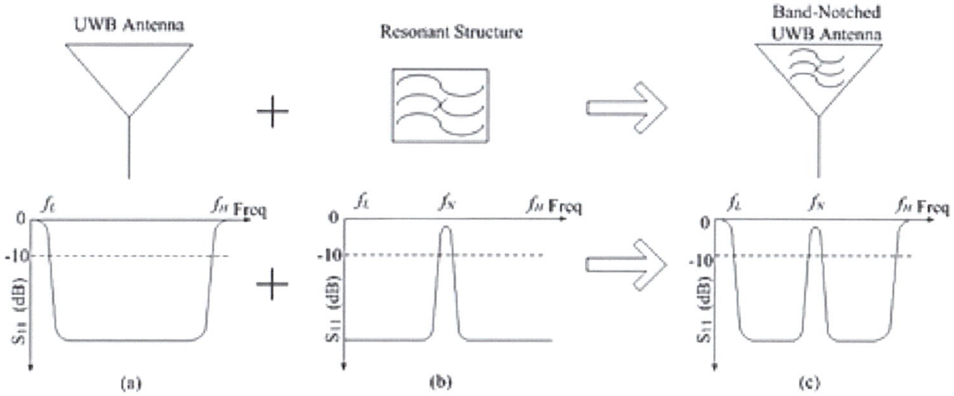

Figure 2.
The basic design of UWB antenna with band-notched characteristic: (a) UWB antenna, (b) Band-notch resonant structure and (c) Band-notched UWB-antenna.

the UWB antenna with band-notched characteristic is shown in **Figure 2** [17, 18]. The basic UWB antenna is having impedance bandwidth from [f_L-f_H] as shown in **Figure 2(a)**, where f_L and f_H are the lowest and highest −10 dB cut-off frequencies of S_{11}, respectively. A band-stop resonant structure also has the impedance bandwidth from [f_L-f_H], but with notch (resonant) frequency at f_N to stop the unwanted frequencies as shown in **Figure 2(b)**. The UWB antenna combining with a band-stop resonant circuit forms band-notched UWB antenna is shown in **Figure 2(c)**. The band-notched UWB antenna does not interfere with existing communication systems which are operating at f_N. Hence, the design of UWB antennas with band-notched function is needed.

The band-notch characteristics for UWB systems can be obtained by integrating band-notch resonator (like slots, split ring resonators, strips and stubs) to the UWB antenna. The possible locations for integration of band-notch resonator are: on or adjacent to radiator or feed line or on the ground as given in **Figure 3**. The length of the band-notch resonator controls the notch center frequency. However, the desired notch band is obtained by proper tuning of length and width of the resonator. The total length of the band-notch resonator should be λ/2 or λ/4 corresponding to the notched-band center frequency as given in equations (1) and (2) [19], where λ is the guided wavelength.

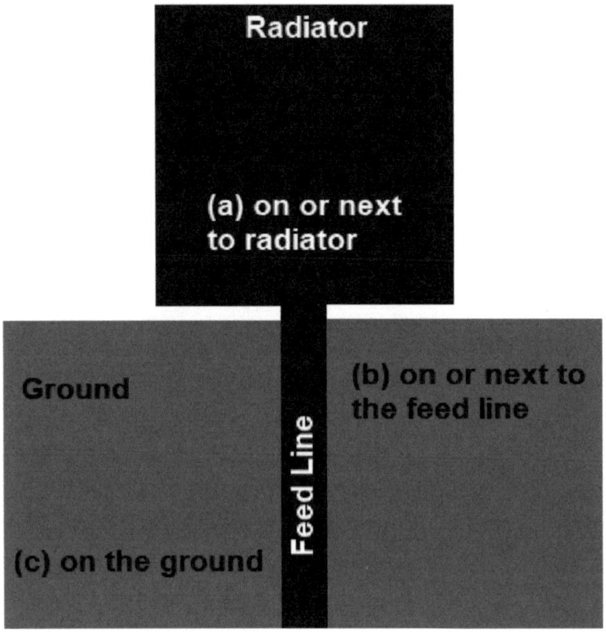

Figure 3.
Possible locations of resonant structure on UWB antenna.

$$L_{Notch} = \frac{c}{2f_{Notch}\sqrt{\varepsilon_{eff}}}, \quad (1)$$

or

$$L_{Notch} = \frac{c}{4f_{Notch}\sqrt{\varepsilon_{eff}}}, \quad (2)$$

$$\lambda = \frac{c}{f_{Notch}\sqrt{\varepsilon_{eff}}}, \quad (3)$$

$$\varepsilon_{eff} = \frac{(1+\varepsilon_r)}{2}, \quad (4)$$

where c denotes light speed, f_{Notch} represents notch center frequency, L_{Notch} is the total length of slot or strip, ε_{eff} indicates effective dielectric constant and ε_r denotes dielectric constant.

Several investigations were reported earlier to create band notch function at WLAN band for ultra-wideband systems in [19–30]. Methods include inserting λ/4 and λ/2 slot resonators on the ground plane [19], using a pair of ground stubs locating along the edge of the ground plane [20], inserting open stub in the printed folded monopole [21], etching folded U-shaped slots in the feed line of the antenna [22, 23], incorporating SRR slots on radiating element [24], quarter-wave stub connected to the ground [25], adding protruding two rectangular stubs on the ground plane [26], with a slot of length 1.0 λ in the radiator [27], open-ended quarter-wavelength L-shaped slots were etched on the rectangular radiating patches [28], using C-shaped and Z-shaped slot resonators on the ground [29], employing elliptical SRR on the radiating element [30].

The antenna designs presented in the above literature exhibiting acceptable isolation and notching characteristics, but some designs were not compact enough and few are a bit complex. So, the design of simple and compact band-notched ultra-wideband MIMO antenna with low mutual coupling is needed. In this chapter, we have presented compact isolation-enhanced planar UWB-MIMO antenna with single band-notched characteristics at WLAN band [31]. Ansoft HFSS v.13 is used to carry out the proposed antenna design, optimization, and simulations. For validating the simulation results, all the proposed antenna has been fabricated and tested using the Agilent N5224A PNA, Anritsu MS2037C vector network analyzer and an anechoic chamber. In Section 2, UWB-MIMO antenna with single band-notched characteristic is discussed. Finally, conclusions of the work are given in Section 3.

2. UWB-MIMO antenna with single band-notched characteristic

Compact ultra-wideband MIMO antenna exhibiting band-notch characteristics at WLAN band (5 to 5.9 GHz) for portable wireless devices is presented in this section [31]. The following sub sections discusses the detailed description of the proposed design.

2.1 Antenna design

The geometry of the proposed single band-notch UWB-MIMO antenna and photograph of the fabricated antenna are shown in **Figure 4(a)** and **(b)**. The design is printed on an FR4 substrate having dielectric constant (ε_r) of 4.4, a thickness of 0.8 mm, and a loss tangent of 0.02. The overall size of the proposed antenna is $L \times W \times h\ mm^3 = 26 \times 40 \times 0.8\ mm^3$. The antenna comprises two identical rectangular planar monopole radiating elements, denoted as PM1 and PM2 having sizes $L_R \times W_R$ as shown in **Figure 4(a)**. Both PM1 and PM2 are fed by the 50-ohm coplanar waveguide having dimensions $F_{L1} \times W_F$. And, the common ground is formed by joining $L_G \times W_G$ and $L_G \times L$ reduced ground planes and is also printed on the same side of the substrate. The planar monopoles PM1 and PM2 are positioned perpendicularly to each other to reduce the mutual coupling between the elements and to improve the isolation between the antenna ports. A long rectangular strip of size $S_L \times S_W$ is extended from the common ground plane between the monopoles to further enhance isolation and improve the impedance bandwidth of the antenna. The ground strip extends the current path which shifts the first resonance frequency to lower band and blocks the surface currents to minimizes the mutual coupling. An inverted U-slot resonator is placed on the feed line to create a band-notch function at 5–5.9 GHz. The antenna optimized dimensions are given as follows: (unit: mm): $D_1 = 5.1$, $D_2 = 6.1$, $D_3 = 11.2$, $F_{L1} = 9.5$, $F_{L2} = 1.5$, $F_{L3} = 0.4$, $L = 26$, $L_G = 8$, $L_R = 10$, $S_L = 18$, $S_W = 1$, $W = 40$, $W_F = 1.8$, $W_G = 3.2$, $W_R = 11$, $U_1 = 7.8$, $U_2 = 0.4$, and $U_W = 0.3$. **Figure 5(a)** and **(b)** shows the simulated S-parameters such as S_{11} and S_{21} of the Antenna 1 (UWB-MIMO antenna without ground strip), Antenna 2 (UWB-MIMO with a ground strip), and the proposed antenna. It can be observed that the proposed ultra-wideband MIMO antenna is operating from 2.2 to 11.4 GHz with good impedance bandwidth except at notch band from 5 to 5.9 GHz. Also, the mutual coupling of less than −20 dB is obtained over the entire UWB band.

2.2 Study of MIMO antenna

Since the ground and radiating elements are having smaller dimensions, the flow of surface currents on the ground plane and near-field radiation leads to poor impedance matching and high mutual coupling, which restricts the performance of MIMO antenna. The ultra-wideband MIMO antenna without and with ground strip

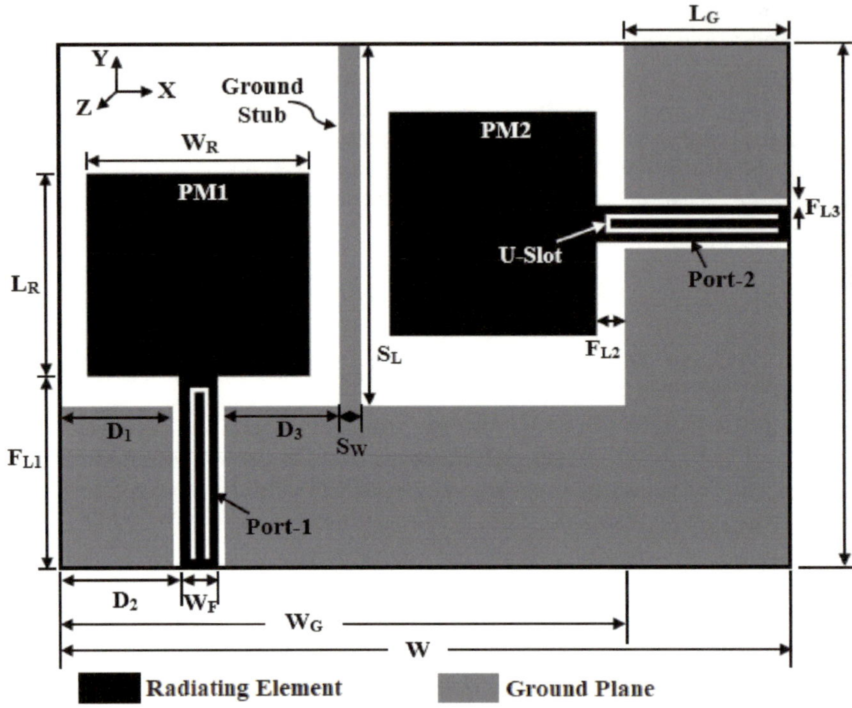

(a)

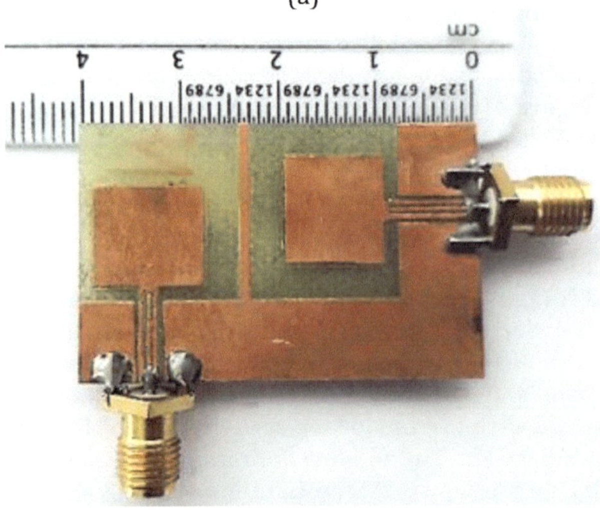

(b)

Figure 4.
(a) Geometry of the proposed antenna and (b) fabricated antenna.

is shown in **Figure 6(a)** and **(b)**, respectively. The effects of the ground strip on impedance bandwidth and mutual coupling between the MIMO antenna elements are plotted in **Figure 7(a)** and **(b)**. With ground strip between the PM1 and PM2 (Antenna 2), the first resonance is generated at 2.5 GHz with a lower cutoff frequency of 2.3 GHz and provides good impedance bandwidth from 2.3 to 11.4 GHz as depicted in **Figure 7(a)**. And, from **Figure 7(b)**, the mutual coupling of lower than −20 dB between the antenna elements is observed throughout the UWB band which is less than −17 dB. In addition, the flow of surface currents is effectively

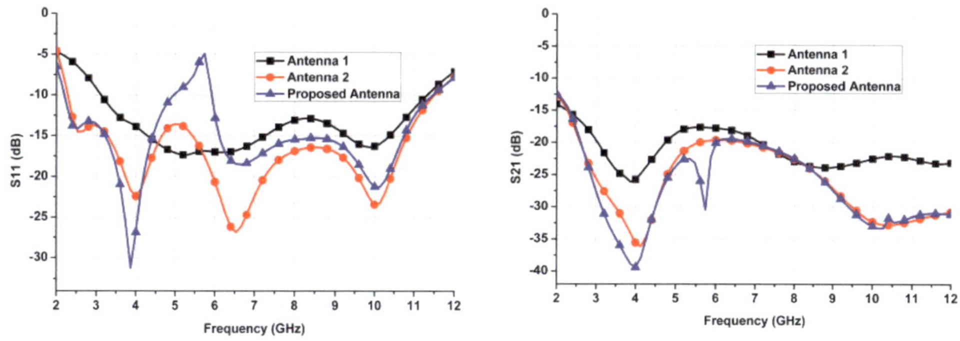

(a) (b)

Figure 5.
Simulated S-parameters. (a) Simulated S11 parameter. (b) Simulated S21 parameter.

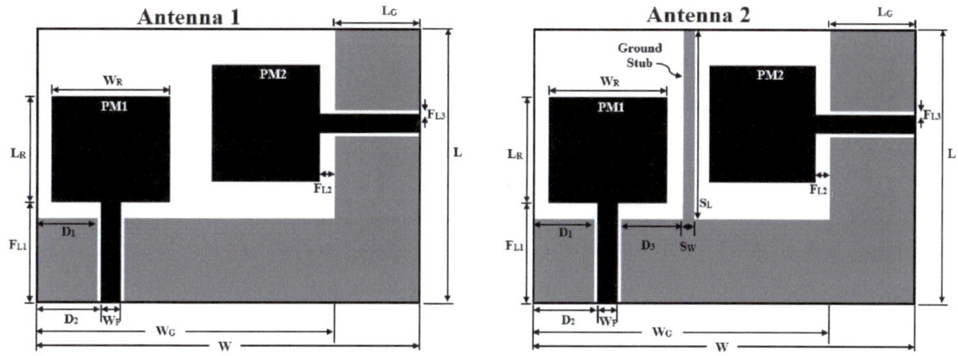

(a) (b)

Figure 6.
(a) UWB-MIMO antenna without a ground strip (Antenna 1), (b) UWB-MIMO antenna with a ground strip (Antenna 2).

suppressed by the ground strip and thus less amount of current is leaked into the port 2 when port 1 is excited as displayed in **Figure 7(c)**. The ground strip can work as a reflecting surface so that the direction of surface currents is diverted and thus the distance between the ports is increased. Hence, the isolation between the MIMO antenna ports is significantly enhanced. Also, the ground strip between the MIMO antenna elements will improve impedance matching characteristics and minimizes the mutual coupling of the MIMO antenna. The MIMO antenna is also studied by varying the ground strip length S_L and width S_W and are plotted in **Figure 8(a)–(d)** and the same tabulated in **Table 1**. It can be observed from that the total length and width of the ground strip has more effect on the impedance bandwidth ($|S_{11}| < -10$ dB) than the isolation or mutual coupling. In this work, the ground strip length S_L = 18 mm and width S_W = 1 mm is adopted.

To create band-notch filtering function for ultra-wideband systems, slots of various shapes or split-ring resonators or strips can be used on or next to the feed line or the radiating element or the ground plane as reported earlier. The slot or SRR or strip can act as a band-notch resonator. The notch band center frequency is controlled by the length of the resonator and notch band bandwidth is controlled by the width of the resonator. In this design, an inverted U-shaped slot is used as a band-notch resonator and is etched on the feed line of Antenna 2 which forms the

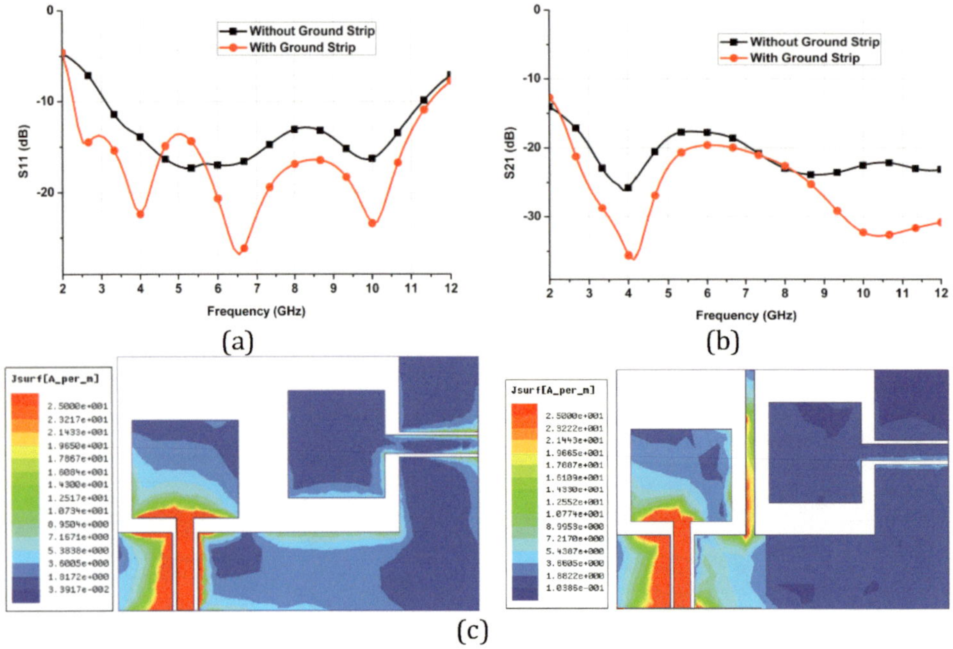

Figure 7.
(a) S11 without and with a ground strip, (b) S21 without and with a ground strip (c) surface current distribution at 3.8 GHz when port 1 excited without and with a ground strip.

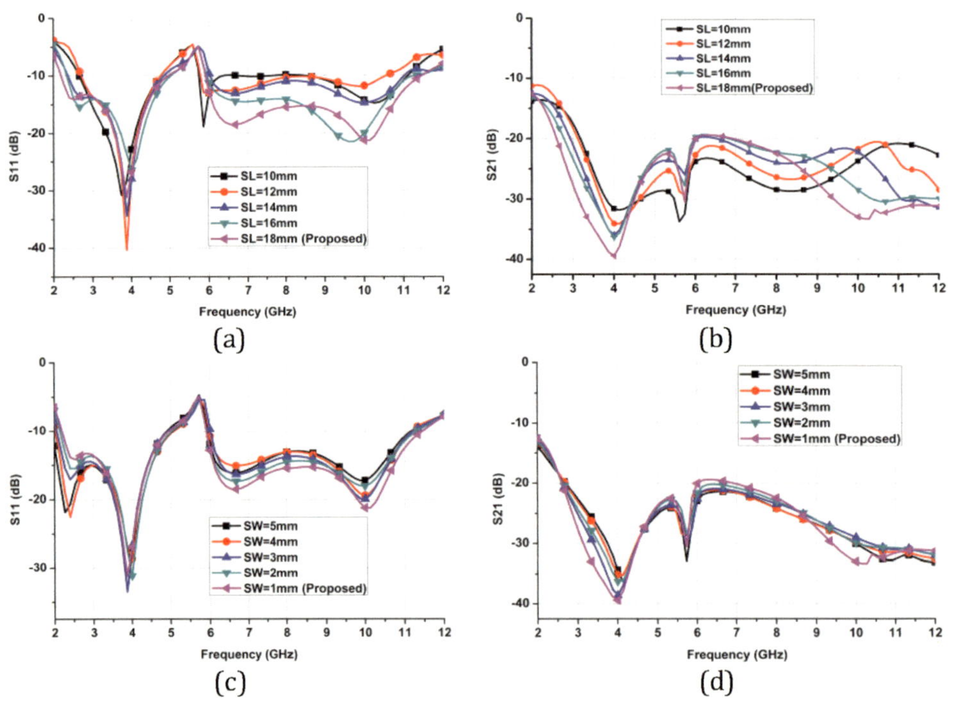

Figure 8.
(a) S11 for different strip lengths SL, (b) S21 different strip lengths SL, (c) S11 for different strip widths SW, (d) S21 for different strip widths SW.

Parameter	Value (mm)	Bandwidth ($S_{11}< -10$ dB)	Mutual coupling ($S_{21} < -20$ dB)
Strip length S_L	10	2.6–6.4	3.3–11.4
	12	2.6–8.6	3.2–11.4
	14	2.4–11.0	2.9–11.4
	16	2.4–11.3	2.8–11.4
	18 (proposed)	2.2–11.4	2.6–11.4
Strip width S_W	5	2.0–11.1	2.8–11.4
	4	2.0–11.1	2.8–11.4
	3	2.1–11.2	2.8–11.4
	2	2.2–11.2	2.7–11.4
	1 (proposed)	2.2–11.4	2.6–11.4

Table 1.
The S_{11} and S_{21} for various strip lengths and widths except at notch band.

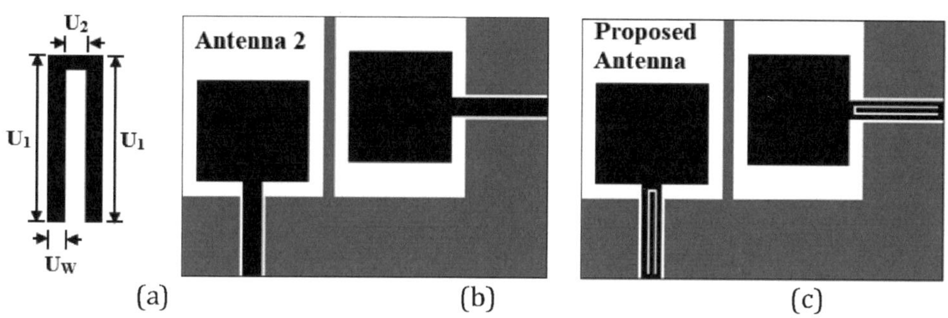

Figure 9.
(a) An inverted U-slot resonator, (b) UWB-MIMO antenna with ground strip (Antenna 2), (c) proposed band-notched UWB-MIMO antenna.

proposed band-notch UWB-MIMO antenna as shown in **Figure 9(a)–(c)**. The length of the U-shaped resonator is calculated using Eq. (5) [19]:

$$L_N = \frac{c}{2f_N\sqrt{\varepsilon_{eff}}} \approx \frac{\lambda}{2}, \tag{5}$$

where L_N denotes the total length of U-slot and f_N is notch center frequency. When f_N = 5.7 GHz and ε_r = 4.4, the calculated length of the U-slot resonator using equation (5) is 16.01 mm. The simulated or designed total length of the inverted U-slot resonator is 16 mm and is determined by using equation (6).

$$L_{U-Slot} = 2\,U_1 + U_2 \approx \frac{\lambda}{2}. \tag{6}$$

Good agreement between the calculated (theoretical) length and simulated (practical) length is observed. **Figure 10(a)–(c)** shows the S_{11}, S_{21} and surface currents without and with inverted U-slot resonator. As seen in **Figure 10(a)**, the proposed antenna is working from 2.2 to 11.4 GHz with good impedance bandwidth and generates band-notch characteristics from 5 to 5.9 GHz with S_{11} of −5 dB at 5.7 GHz. And the mutual coupling of below −20 dB over the entire working band is observed as from the **Figure 10(b)**. It is evident from **Figure 10(c)** that at 5.7 GHz,

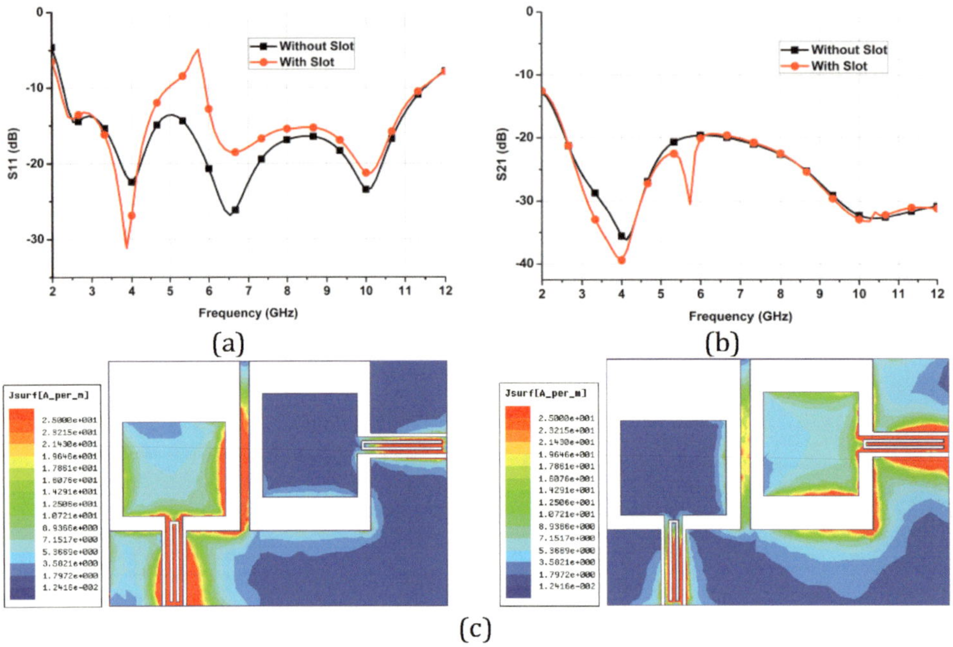

Figure 10.
(a) S11 without and with inverted U-slot, (b) S21 without and with inverted U-slot (c) current distribution at 5.7 GHz when port 1 and port 2 excited.

Parameter	Value (mm)	Notch-band (GHz)	Notch-center frequency f_N (GHz)
Slot length U_1	6.8	6.26–6.7	6.5
	7.3	5.73–6.4	6.2
	7.8 (proposed)	5.0–5.9	5.7
	8.1	4.8–5.7	5.4
Slot width U_W	0.25	5.2–5.7	5.6
	0.3 (proposed)	5.0–5.9	5.7
	0.35	4.95–6	5.7
	0.4	4.94–6.1	5.7

Table 2.
The notch bands and notch center frequencies for different slot lengths and widths.

heavy current is concentrated around the inverted U-slot resonator which acts a band-notch filter, so the current flow on the radiating elements is blocked and hence no radiation from the antenna. Therefore, notch band from 5 to 5.9 GHz WLAN band is created.

The parametric analysis on the slot length U_1 and slot width U_W is performed to describe the effects of inverted U-slot. **Table 2** shows the notch bands and notch center frequencies for different slot lengths and widths. **Figure 11(a)** and **(b)** illustrates the S_{11} of the MIMO antenna for different slot lengths U_1 and slot widths U_W, respectively. It is evident that as the slot length U_1 increasing from 6.8 mm to 8.1 mm, the center frequency of notch f_N is decreasing from 6.5 GHz to 5.4 GHz and notch band is shifting from (6.26–6.7) GHz to (4.8–5.7) GHz. The required band notch from 5 to 5.9 GHz is generated for U_1 of 7.8 mm which is used in this design. Form **Figure 11(b)**, it can be observed that increasing the slot width U_W from 0.25

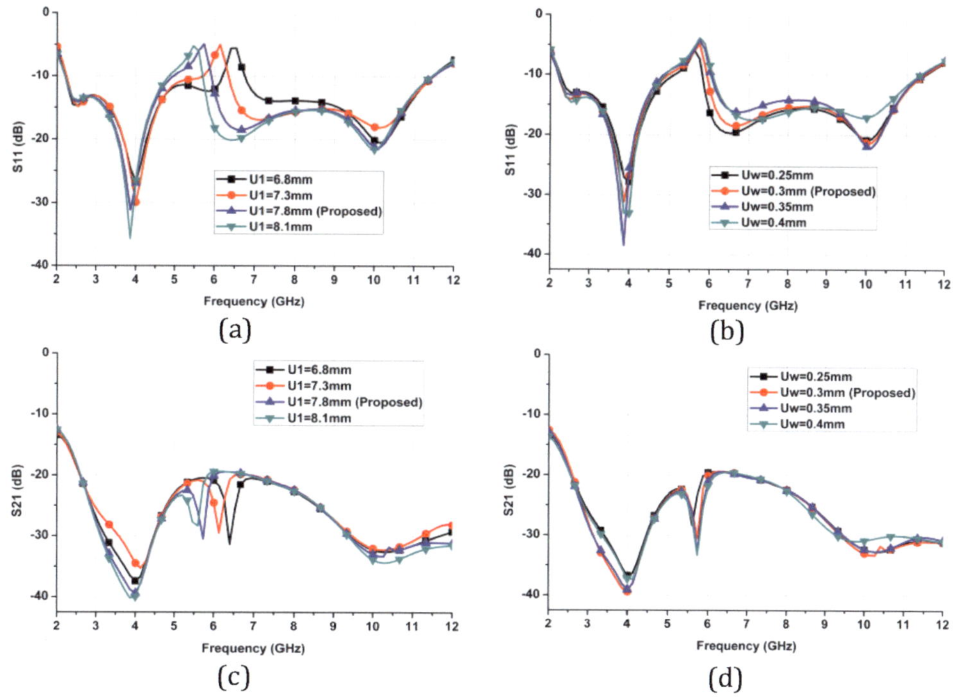

Figure 11.
S11-parameter for (a) various slot lengths, (b) various slot widths; and S21 for (c) various slot lengths, (d) various slot widths.

mm to 0.4 mm, the notch bandwidth is increasing from (5.2–5.7) GHz to (4.94–6.1) GHz with notch center frequency f_N at 5.7 GHz. The slot width U_W of 0.3 mm is chosen in this proposed design to get the desired band notch from 5 to 5.9 GHz. **Figure 11(c)** and **(d)** shows the effects of inverted U-slot resonator on the S_{21} of the MIMO antenna for various slot lengths and widths. Also, it is observed that variation in the slot length U_1 and width U_W has a negligible effect on the mutual coupling of MIMO antenna.

2.3 Results and discussion

The proposed antenna offers good impedance bandwidth ($|S_{11}|<-10$ dB) from 2.2 to 11.4 GHz with band notch at 5–5.9 GHz as demonstrated in **Figure 12(a)**. Hence, the frequency interference from WLAN band can be effectively suppressed by the proposed UWB MIMO antenna. And, from **Figure 12(b)**, it is found that the simulated and measured mutual coupling (S_{21}) value is about −20 dB in the operating band except at few frequencies around 6.2 and 7.2 GHz (−18 dB) demonstrating good isolation between the ports. At 6.2 and 7.2 GHz the S_{21} is −16 dB. From **Figure 12(c)**, the antenna has good 2:1 VSWR from 2.2 to 11.4 GHz excluding at notch-band, i.e. from 5.0 to 5.9 GHz. It is observed that the VSWR of about 3.5 at 5.7 GHz. The summary of simulated and mesured results presented in **Figure 12** are provided in **Table 3**. The simulated and measured radiation patterns of the proposed antenna on the *E*-plane and *H*-plane at 3.8, 6.5, and 10 GHz when port 1 is excited and port 2 is terminated with 50-ohm load, and vice-versa are shown in **Figure 13(a)** and **(b)**. Good agreement between the simulated and measure 2-D radiation patterns is observed. At 3.8 and 6.5 GHz frequencies, PM1 and PM2 have quite omnidirectional radiation patterns in H-planes, i.e. the *XZ* plane and the *YZ* plane, respectively. However, at 10 GHz because of the higher-order resonances,

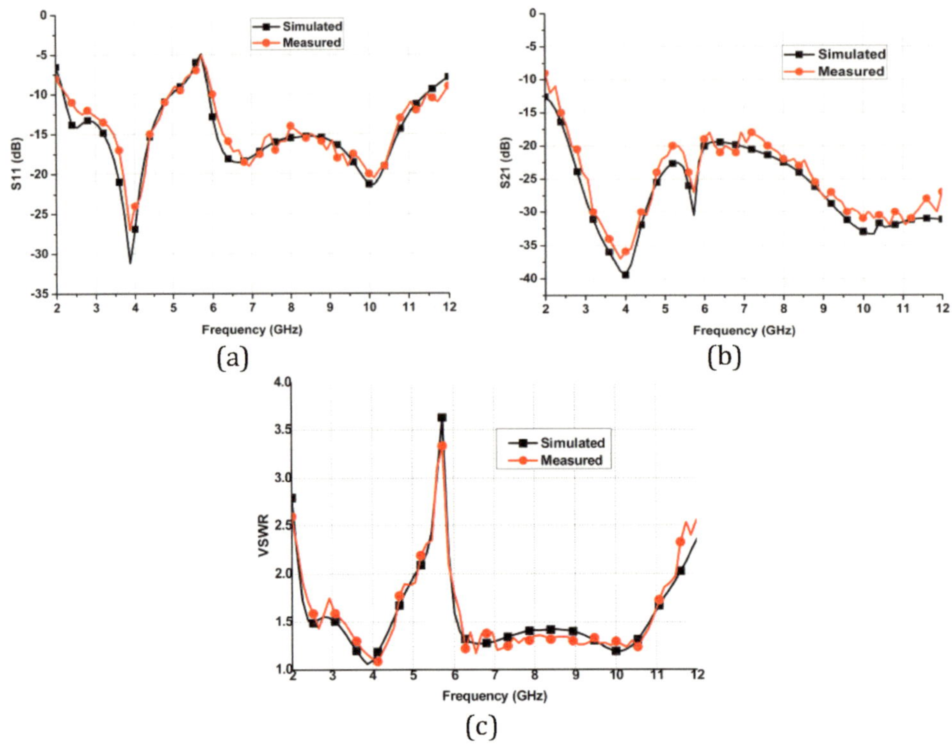

Figure 12.
The simulated and measured results: (a) S11-parameter, (b) S21-parameter and (c) VSWR

Result	Bandwidth ($S_{11} < -10$ dB)	Mutual coupling (S_{21})	VSWR at notch band	Notch band
Simulated	2.2–11.4 GHz	<−20 dB	3.6	5–5.9 GHz
Measured	2.3–11.7 GHz	<−18 dB	3.4	5–6 GHz

Table 3.
The summary of simulated and measured results presented in ***Figure 12****.*

the radiation patterns in the *H*-planes are less omnidirectional. And, at 6.5 and 10 GHz, PM1 and PM2 have the dumbbell-shaped or bidirectional patterns in the *E*-planes, i.e. the *YZ* plane and the *XZ* plane, respectively. However, at 3.8 GHz, PM1 and PM2 do not have the "dumb-bell" shaped patterns in the *E*-planes, because, the strip on the ground plane changes the current distributions. It can be seen from **Figure 13** that the proposed antenna provides omnidirectional radiations in *H*-plane which is essential for portable wireless devices to receive the signals from all directions. And, it is also found that *H*-plane patterns of port 1 and port 2 are nearly mirror images demonstrating the good pattern diversity. The simulated and measured peak gain of the proposed design is plotted in **Figure 14(a)**. The peak gain of 2.4 to 7.5 dBi in the operating band is observed excepting at the notch band. At the notch band, the measured peak gain falls to −2.2 dBi. **Figure 14(b)** shows the simulated and measured radiation efficiency plot of the proposed antenna. The radiation efficiency of above 90% is found across the UWB band excluding at 5–5.9 GHz notch band. At notch band, the efficiency drops to 12%. It is evident from **Figure 14(a)** and **(b)** that the proposed antenna can avoid the frequency interference from WLAN band more efficiently.

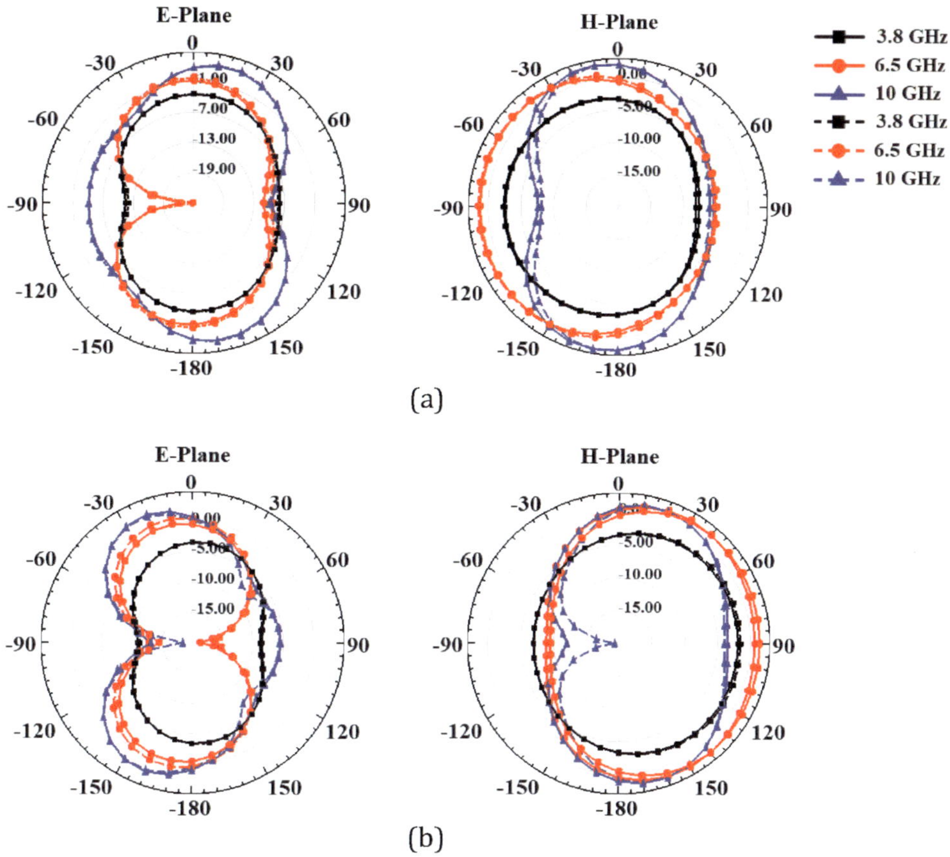

Figure 13.
Simulated (solid line) and measured (dashed line) radiation patterns. (a) simulated and measured when port 1 excited. (b) Simulated and measured when port 2 excited.

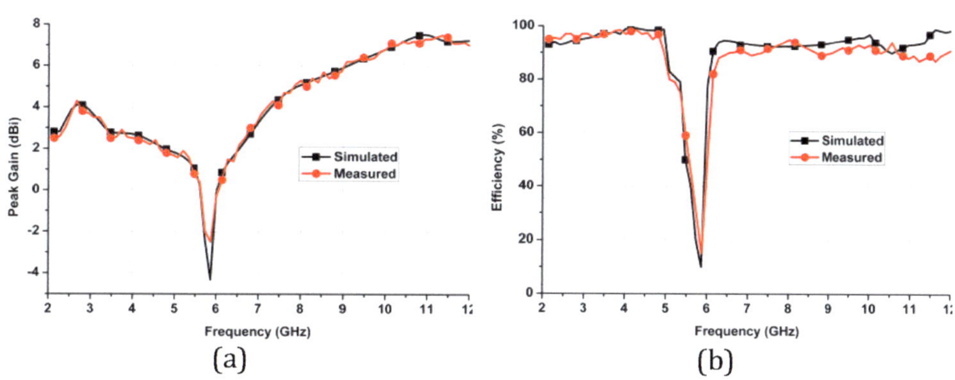

Figure 14.
Simulated and measured (a) peak gain, (b) radiation efficiency.

Along with the radiation patterns, envelope correlation coefficient (ECC) is also an important parameter to study the MIMO antenna diversity and is calculated using *S*-parameters with the equation (7) of a two-port MIMO antenna system reported by Blanch et al. [32]. The envelop correlation coefficient (ECC) measures the similarity between the antenna radiation patterns and is very useful to estimate the performance of MIMO antenna. The lower the ECC value means the lesser is the

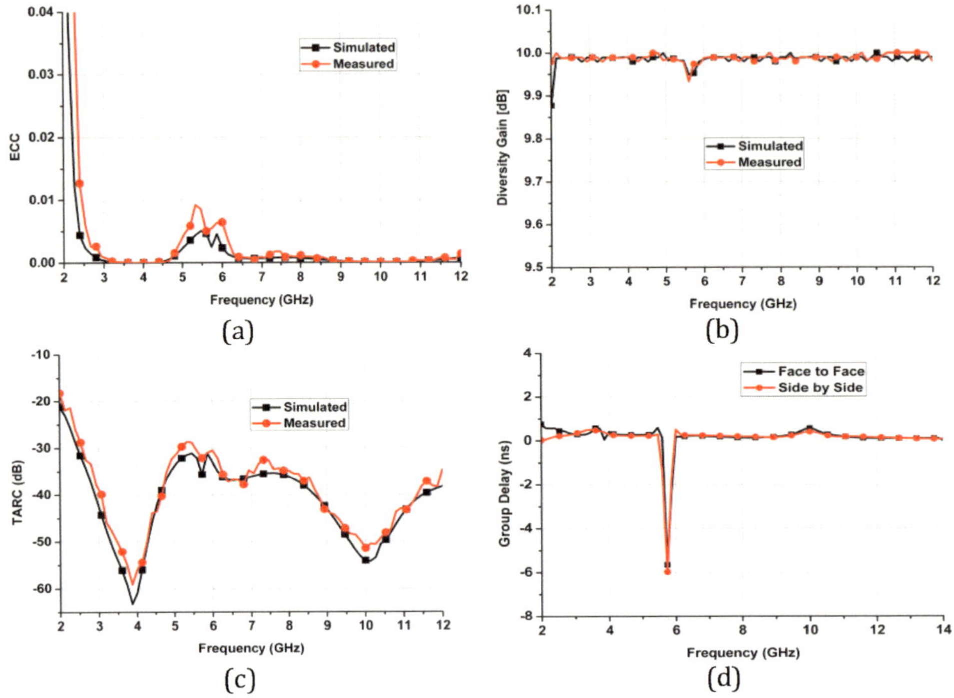

Figure 15.
(a) Simulated and measured ECC, (b) Simulated and measured diversity gain, (c) Simulated and measured TARC, (d) Group delay of the proposed antenna.

overlapping between the two radiation patterns. For MIMO antenna system to ensure the diversity performance as good, the ECC with value below 0.5 is adopted in most the cases. **Figure 15(a)** shows the simulated and measured ECC of the proposed antenna. The simulated ECC is about 0.005 and measured ECC is below 0.008 from 2.2 to 11.4 GHz.

$$ECC = \frac{\left|S_{11}^{*}S_{12} + S_{21}^{*}S_{22}\right|^2}{\left(1 - \left(|S_{11}|^2 + |S_{21}|^2\right)\right)\left(1 - \left(|S_{22}|^2 + |S_{12}|^2\right)\right)}, \tag{7}$$

The diversity gain (DG) and total active reflection coefficient (TARC) are also essential parameters to study the MIMO antenna diversity performance. The diversity gain and total active reflection coefficient of the proposed antennas can be estimated by using the equations (8) and (9) [33] as follows:

$$DG = 10\sqrt{1 - ECC^2} \tag{8}$$

$$TARC = \sqrt{\frac{(S_{11} + S_{12})^2 + (S_{21} + S_{22})^2}{2}} \tag{9}$$

The simulated and measured diversity gain plots are given in **Figure 15(b)**. The diversity gain of >9.95 dB is found in the UWB band. And, **Figure 15(c)** shows the simulated and measured TARC. It is observed that the TARC of less than −28 dB is obtained in the whole UWB band. The group delay of the proposed antenna is measured in face to face and side by side situations with the space of 30 cm is shown in **Figure 15(d)**. The group delay is almost uniform and is below 1 ns in the complete working band except at stopband. At the notch band i.e. at 5.7 GHz, the group delay of 5.8 ns in the face to face orientation and 6 ns in the side by side orientation ensures that the proposed antenna can transmit the UWB signal with minimum distortion. It is observed from the above results that there is good agreement between the simulated and measured S-parameters except for some deviations due to fabrication and soldering imperfections, losses in dielectrics and conductors, effects of SMA connector, and measurement tolerances.

3. Conclusion

To mitigate the frequency interference from narrowband system like WLAN, a compact planar UWB antenna with single band-notched characteristics for portable wireless devices applications is discussed in this chapter. In this design, the monopoles are arranged perpendicularly to reduce to mutual coupling. A rectangular strip is extended from the ground plane to improve the impedance matching characteristics and to reduce the mutual coupling further or enhance the isolation. An inverted U-shaped slot is used on the feed line to realize the band-notch filtering function for suppressing the frequency interference from 5 to 5.9 GHz WLAN band. For validating the simulation results, all the proposed antennas have been fabricated and tested using the Agilent N5224A PNA, Anritsu MS2037C vector network analyzer and an anechoic chamber. The measured results of all the proposed antenna are well agreed with simulated results. The measured and simulated results show that the proposed antenna offer good impedance bandwidth of $S_{11} \leq -10$ dB in whole UWB band (3.1–10.6 GHz) except at the designed notch bands while giving less mutual coupling (S_{21}) of lower than -20 dB in the entire UWB band. The low envelope correlation coefficient, nearly constant gain, stable radiation patterns, more directive gain, TARC and less group delay, demonstrate that the proposed MIMO antenna is an appropriate choice for portable wireless UWB systems.

Acknowledgements

Authors would like to express their gratitude towards University College of Engineering & Technology, Acharya Nagarjuna University, Guntur and management of Koneru Lakshmaiah Education Foundation, Guntur for their continuous support and encouragement during this work. Further, Dr. J. Chandrasekhar Rao and Dr. N. Venkateswara Rao would like to acknowledge with thanks DST through FIST grant SR/FST/ETI-316/2012, ECR/2016/000569 and Dr. B.T.P. Madhav, Prof. of ECE, KLEF for providing measurement facility in LCRC lab. I would like to thank Mr. P. Ramakoti Reddy, Electro Circuit Systems, Hyderabad, Mr. K. Vijaya Saradhi Reddy, Excel Radio Frequency Technologies, Hyderabad and Mr. Krishna Prasad, Scientist-ECIL, Hyderabad for their help in the fabrication and measurements of the prototype developed in this work.

Author details

Chandrasekhar Rao Jetti* and Venkateswara Rao Nandanavanam
Bapatla Engineering College, Bapatla, Andhra Pradesh, India

*Address all correspondence to: jettychandu@gmail.com

References

[1] Federal Communications Commission (FCC), Revision of Part 15 of the Commission's Rules Regarding Ultra-Wideband Transmission Systems First Rep. and Order, ET Docket 98-153, FCC 02-48, Adopted: Feb. 2002; Released, Apr. 2002.

[2] Kaiser T, Zheng F, Dimitrov E. An overview of ultra-wide-band systems with MIMO. Proceedings of the IEEE. 2009 Feb 27;97(2):285-312.

[3] Mabrouk IB, Talbi L, Nedil M, Hettak K. MIMO-UWB channel characterization within an underground mine gallery. IEEE Transactions on Antennas and Propagation. 2012 Jul 10; 60(10):4866-74.

[4] Tran VP, Sibille A. Spatial multiplexing in UWB MIMO communications. Electronics letters. 2006 Aug 3;42(16):1.

[5] Zheng L, Tse DN. Diversity and multiplexing: A fundamental tradeoff in multiple-antenna channels. IEEE Transactions on information theory. 2003 May 7;49(5):1073-96.

[6] Liu L, Cheung SW, Yuk TI. Compact MIMO antenna for portable devices in UWB applications. IEEE Transactions on antennas and propagation. 2013 May 15;61(8):4257-64.

[7] Zhang S, Ying Z, Xiong J, He S. Ultrawideband MIMO/diversity antennas with a tree-like structure to enhance wideband isolation. IEEE Antennas and Wireless Propagation Letters. 2009 Nov 24;8:1279-82.

[8] Luo CM, Hong JS, Zhong LL. Isolation enhancement of a very compact UWB-MIMO slot antenna with two defected ground structures. IEEE Antennas and Wireless Propagation Letters. 2015 Apr 15;14: 1766-9.

[9] Tao J, Feng Q. Compact ultrawideband MIMO antenna with half-slot structure. IEEE Antennas and wireless Propagation letters. 2016 Aug 30;16:792-5.

[10] Gallo M, Antonino-Daviu E, Ferrando-Bataller M, Bozzetti M, Molina-Garcia-Pardo JM, Juan-Llacer L. A broadband pattern diversity annular slot antenna. IEEE Transactions on Antennas and propagation. 2011 Dec 20; 60(3):1596-600.

[11] Rao JC, Rao NV. CPW-fed compact ultra wideband MIMO antenna for portable devices. Indian Journal of Science and Technology. 2016 May;9 (17):1-9.

[12] Bassi M, Caruso M, Khan MS, Bevilacqua A, Capobianco AD, Neviani A. An integrated microwave imaging radar with planar antennas for breast cancer detection. IEEE Transactions on microwave theory and techniques. 2013 Feb 26;61(5): 2108-18.

[13] Capobianco AD, Khan MS, Caruso M, Bevilacqua A. 3–18 GHz compact planar antenna for short-range radar imaging. Electronics letters. 2014 Jun 27;50(14):1016-8.

[14] Zhang S, Pedersen GF. Mutual coupling reduction for UWB MIMO antennas with a wideband neutralization line. IEEE antennas and wireless propagation letters. 2015 May 21;15:166-9.

[15] Li H, Liu J, Wang Z, Yin YZ. Compact 1×2 and 2×2 MIMO antennas with enhanced isolation for ultrawideband application. Progress In Electromagnetics Research. 2017;71: 41-9.

[16] Kerkhoff A, Ling H. Design of a planar monopole antenna for use with

ultra-wideband (UWB) having a band-notched characteristic. InIEEE Antennas and Propagation Society International Symposium. Digest. Held in conjunction with: USNC/CNC/URSI North American Radio Sci. Meeting (Cat. No. 03CH37450) 2003 Jun 22 (Vol. 1, pp. 830-833). IEEE.

[17] Weng YF, Cheung SW, Yuk TI, Liu L. Creating band-notched characteristics for compact UWB monopole antennas. InUltra Wideband-Current Status and Future Trends 2012 Oct 3. IntechOpen.

[18] Liu L. Compact planar UWB antennas for wireless device applications. HKU Theses Online (HKUTO). 2014.

[19] Zheng ZA, Chu QX, Tu ZH. Compact Band-Rejected Ultrawideband Slot Antennas Inserting With λ/2 and λ/4 Resonators. IEEE Transactions on Antennas and Propagation. 2011 Feb;59 (2):390-7.

[20] Weng YF, Cheung SW, Yuk TI. Compact ultra-wideband antennas with single band-notched characteristic using simple ground stubs. Microwave and Optical Technology Letters. 2011 Mar;53 (3):523-9.

[21] Lee JM, Kim KB, Ryu HK, Woo JM. A compact ultrawideband MIMO antenna with WLAN band-rejected operation for mobile devices. IEEE Antennas and wireless propagation letters. 2012 Aug 21;11:990-3.

[22] Sayidmarie KH, Najm TA. Performance evaluation of band notch techniques for printed dual band monopole antennas. International Journal of Electromagnetics and Applications. 2013;3(4):70-80.

[23] Majeed AH, Abdullah AS, Sayidmarie KH, Abd-Alhameed RA, Elmegri F, Noras JM. Compact dielectric resonator antenna with band-notched characteristics for ultra-wideband applications. Progress In Electromagnetics Research. 2015;57: 137-48.

[24] Gao P, He S, Wei X, Xu Z, Wang N, Zheng Y. Compact printed UWB diversity slot antenna with 5.5-GHz band-notched characteristics. IEEE Antennas and Wireless Propagation Letters. 2014 Feb 14;13:376-9.

[25] Khan MS, Capobianco AD, Naqvi A, Shafique MF, Ijaz B, Braaten BD. Compact planar UWB MIMO antenna with on-demand WLAN rejection. Electronics Letters. 2015 Jun 3;51(13): 963-4.

[26] Liu L, Cheung SW, Yuk TI. Compact MIMO antenna for portable UWB applications with band-notched characteristic. IEEE Transactions on Antennas and Propagation. 2015 Feb 24; 63(5):1917-24.

[27] Sipal D, Abegaonkar MP, Koul SK. Compact band-notched UWB antenna for MIMO applications in portable wireless devices. Microwave and Optical Technology Letters. 2016 Jun;58(6): 1390-4.

[28] Tao J, Feng QY. Compact isolation-enhanced UWB MIMO antenna with band-notch character. Journal of ElEctromagnEtic WavEs and applications. 2016 Nov 1;30(16): 2206-14.

[29] Liu Z, Wu X, Zhang Y, Ye P, Ding Z, Hu C. Very compact 5.5 GHz band-notched UWB-MIMO antennas with high isolation. Progress In Electromagnetics Research. 2017;76: 109-18.

[30] Tripathi S, Mohan A, Yadav SK. A compact MIMO/diversity antenna with WLAN band-notch characteristics for portable UWB applications. Progress In Electromagnetics Research. 2017;77: 29-38.

[31] Jetti CR, Nandanavanam VR. Compact MIMO antenna with WLAN band-notch characteristics for portable UWB systems. Progress In Electromagnetics Research. 2018;88: 1-12.

[32] Blanch S, Romeu J, Corbella I. Exact representation of antenna system diversity performance from input parameter description. Electronics letters. 2003 May 1;39(9):705-7.

[33] Najam AI, Duroc Y, Tedjini S. Multiple-input multiple-output antennas for ultra wideband communications. IntechOpen. 2012 Oct 1;10:209-36.

Section 3

Applications of Ultra-Wideband Technologies

Chapter 4

Medical Application of Ultra-Wideband Technology

Abdulhameed Habeeb Alghanimi

Abstract

This chapter deals with the applications of ultra-wideband technology, especially for medical scope, and the most features and advantages that made it useful in this scope. Also, the chapter has been included with the most important medical applications of UWB technology. Ultra-wideband radar for angiography and UWB glucometer are the main applications which will be explained in this chapter. The exposure for safety aspects, the dielectric properties of human tissues, blood dielectric properties measurement using open-ended coaxial probe experiment to improve the blood image, and the ideal ultra-wideband pulses' shape, width, and repetition time that are used for medical applications have been illustrated. Finally, the results (figures, tables, and experiment results), conclusions, and discussions have been mentioned.

Keywords: UWB medical application, UWB medical radar, UWB glucometer, UWB features, UWB exposure safety aspects

1. Introduction

UWB is an emerging wireless communication technology that introduces a wide approach of wireless techniques in various disciplines, especially in medical applications. This technology works with a frequency range of 3.1–10.6 GHz and power spectrum density (PSD) of −41.3 dBm/MHz according to the American Federal Communications Commission (FCC) as depicted in **Figure 1** [1] and International Commission on Non-Ionizing Radiation Protection (ICNIRP) safety guidelines [2]. This frequency range with low-power consumption makes the technology suitable for use in medical applications. It has no biological side effects and has nonionizing radiation (only thermal effect) as well as it has a good ability to penetrate the human tissues. These features encourage the researchers to propose many studies that have invested UWB in medical applications; most of these papers would focus either on the differences in the dielectric properties of human tissue like breast cancer detection, or on the organ movement detection like heart rate and respiratory detection. The main problems that have been faced by such researches are the absorption and attenuation of the signal by the skin and the vicinity layers, while the returning signal from the deep layers is very weak, as well as the inability to distinguish between the tissues that have convergent dielectric properties. These problems will introduce new challenges to be solved by researchers.

1.1 UWB advantages and features in medical application

- Good ability of penetration for human tissue.

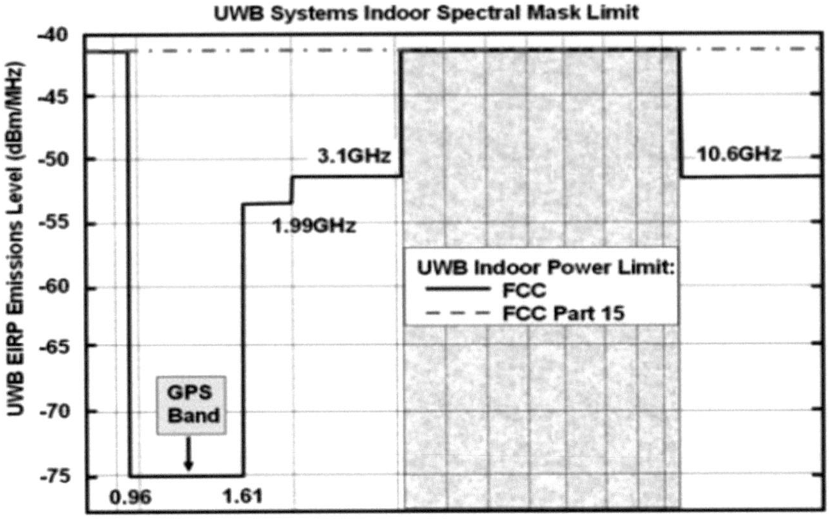

Figure 1.
Indoor power spectrum mask from FCC [1].

- Selective addressing (multiuser).
- High capacity of channel.
- Low cost, low-power consumption, and low complexity.
- Noise-like signal.
- Low probability of interception, jamming, and resistive to a multipath problem.
- High resolution in the time domain making UWB used for location and tracking applications.
- Do not have any biological side effects on the human tissue (low power and nonionizing).

1.2 UWB monocycle pulses

UWB medical radar naturally deals with human body tissues, which will absorb and affect on the radiated energy of UWB pulses, so we have a challenge with increasing the radiated energy without crossing the FCC mask [1], which depends on the Gaussian pulse shape (derivative order), width, and repetition time (frequency). Here, the first derivative of Gaussian pulse equation is:

$$\mathbf{Gs}(\mathbf{t}) = \mathbf{A}\ \exp\left[-\left(\frac{t}{\tau}\right)^2\right] \tag{1}$$

and the fifth derivative of Gaussian pulses is as the following as in Eq. (2), and the frequency range of indoor application from 4 to 6 GHz introduces the best performance and best masking to FCC [1, 3]:

$$\mathbf{Gs}(\mathbf{t}) = \mathrm{A}\left(-\frac{t^5}{\sqrt{2\pi}\tau^{11}} + \frac{10t^3}{\sqrt{2\pi}\tau^9} + \frac{15t}{\sqrt{2\pi}\tau^7}\right)\exp\left(-\frac{t^2}{2\tau^2}\right) \tag{2}$$

where A is the pulse amplitude, t is time, and τ is a time constant as illustrated in **Figure 2** [4]. So, it has safe electromagnetic field according to FCC and the *International Commission on Non-Ionizing Radiation Protection (*ICNIRP) guidelines and it may be causing thermal effects related to the power absorption by human tissue [1, 2, 5].

1.3 The dielectric properties

The dielectric properties are the fundamental parameters that affect the propagation of the electric field. It is a measure of how electric field behaves or interacts with materials, which can be used (for example) to understand how easily an electric field will polarize a given dielectric material. Dielectric constant and loss tangent are both numerical values using which permittivity of a dielectric material can be defined. And the conductivity is the extent of electric current that flow through it. Where the conductivity is used for the rate or degree that electromagnetic wave, electricity, heat, or sound travel through a certain medium.

The dielectric properties of blood have been affected by many coefficients like blood temperature [6], applied electromagnetic wave frequency, clotting rate, human gender [7], blood group type (A, B, AB, and O) [8], blood composition, blood hematocrit level, and hemoglobin percentage [9].

1.4 Blood dielectric properties measurement using open-ended coaxial probe

The dielectric properties will be different from one material to another where these differences will enable the recognizing of the tissues by recognizing its dielectric properties [10]. This experiment attempts to improve the blood dielectric properties individually for increasing the blood appearing over other substances (perfect blood image). This microwave measurement method (experiment) has been applied in 5 GHz frequency center and in a temperature of 37°C; the blood samples with different additions and concentrations have been tested and the results have been recorded. The required experiment devices are shown in **Figure 3**.

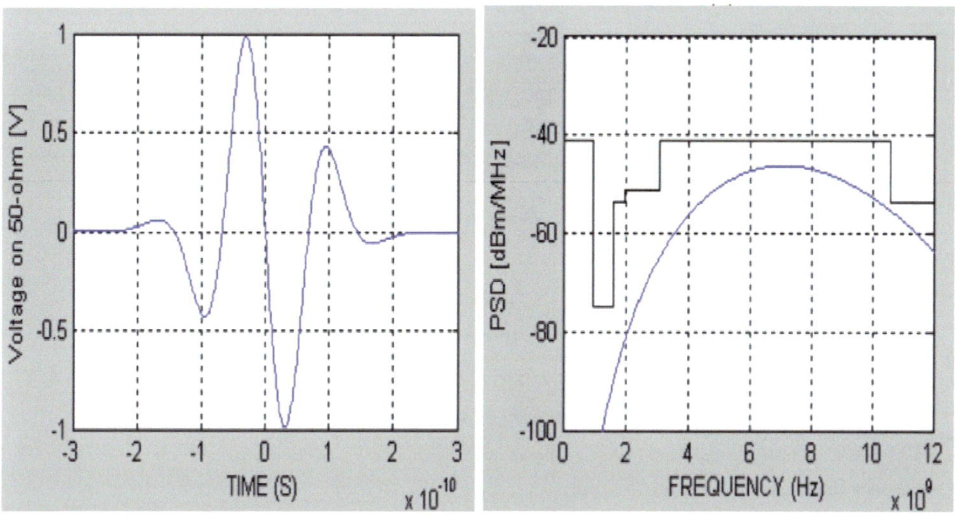

Figure 2.
Fifth derivative Gaussian pulse [4].

The measurement system has the following components:

- RF vector network analyzer (compatible with the above probe model: E5063A ENA series).
- Open-ended coaxial probe (model: 85070E performance probe).
- Water bath and thermistor.
- Adjustable probe stand.
- PC computer (laptop).
- Glass sample containers and alcohol wipes.

The permittivity equation will be as follow [11]:

$$\varepsilon = \frac{2\Gamma \sin\left[2s + \frac{2\pi(L_2 - L_1)}{\lambda}\right.}{s\left\{1 + \Gamma^2 + 2\Gamma \cos\left[2s + \frac{2\pi(L_2 - L_1)}{\lambda}\right]\right)} \tag{3}$$

where λ is the wavelength, Γ is the reflection coefficient, and s is standing wave ratio.

1.5 Safety aspects for the human tissue that exposed to UWB

The use of UWB electromagnetic wave spectrum in medical application that penetrates human body tissue makes a challenge with the patient safety (safe exposure). On the other hand, because its band is 3.1–10.6 GHz, the major effects of UWB using on the human tissue are the thermal effects. These effects are caused by

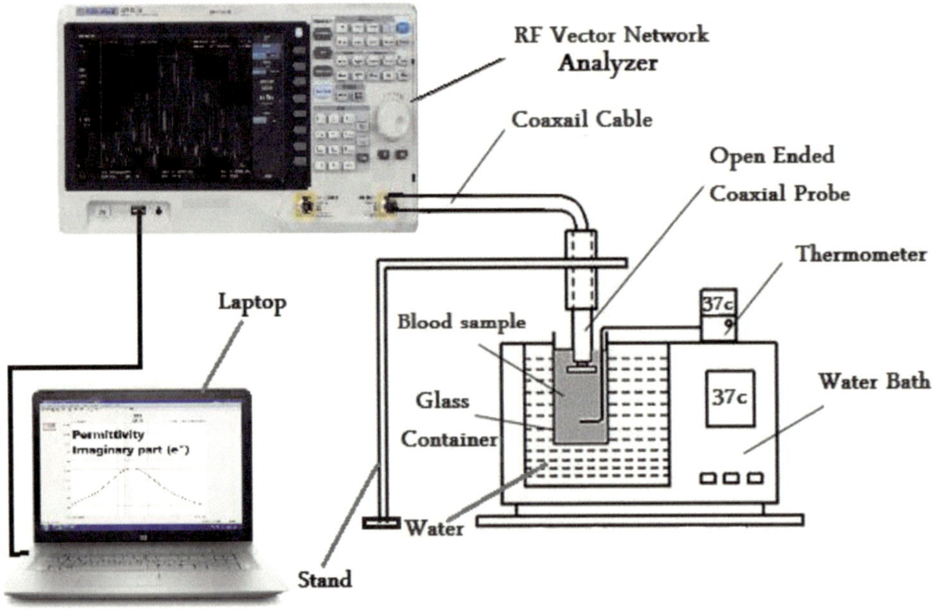

Figure 3.
Blood dielectric properties measurement system.

power absorption and related to the specific absorption rate (SAR) [12], which is measured by W/kg. The available research indicates an increase in the SAR level of whole body equals to 1 and 4 W/kg when it exposes to an electromagnetic field up to 100 kHz for about 30 min, and this increase in SAR causes an increase in the temperature of body less than 1°C [2]. Also, the animal data prove that if the SAR level is raised over than 4 W/kg, it can be out of body control and cause harmful effects of tissue heating, while epidemiological surveys are showing that no biological effects are indicated on workers or the public in the same environment [2].

The SAR level is limited by ICNIRP [2] and FCC [1]. The basic restrictions' limitation is 0.4 W/kg for the worker, 10 W/kg averaged over 10 g mass for head and spinal zones, and 0.08 W/kg for the public population. Moreover, "although little information is available on the relation between biological effects and peak values of pulsed fields, it is suggested that for frequencies exceeding 10 MHz, the power density as averaged over the pulse width should not exceed 1000 times the reference levels or that field strengths should not exceed 32 times the field strength reference levels" [13].

2. UWB medical application

Many papers have been published by many researchers and organizations that have proposed ultra-wideband for medical applications, with different frequencies and hypothesis like:

- Breast tumor detection.
- Bone cancer detection.
- Brain hemorrhage detection.
- Position and localization.
- Noncontacting medical imaging.
- Heartbeat and lung movement detection.
- Detection of vascular pressure.
- Vital sign monitoring.
- MRI image improvement.
- Heart volume detection.
- Ultra-wideband radar for angiography.
- Blood glucose concentration level measurement.

Some of these researches are mentioned below:

In 2002, Staderini [14] presented biomedical applications of UWB radar as a mix of ordinary radar (ranging and detection) with spread spectrum radio which combines the two technologies. The paper talked the problem and interaction of radar wave energy with human tissues, and also the motion of internal organs of body and noncontact probe. The flaw of this paper is the use of 1500 MHz frequency, which is out of the UWB range (3.1–10.6 GHz). So, the results were not intrinsically right.

In 2005, Paulson et al. [15] introduced an overview about the ability of UWB sensor to monitor the internal organs, sense the respiratory, and detect the cardiac function. The noncontacting image with a micro-power impulse remote sensing and low complexity has been introduced, with examples for applying UWB in medical applications. The drawback of this paper is the use of 2 GHz frequency, which is also out of the UWB range (3.1–10.6 GHz). So, the results were not intrinsically right.

In 2007, Staderini et al. [16] proposed an optimal pulsed UWB medical radar for heart beat detection, by detecting the tracking of heart wall movement depending on the obtained pulse echo average and power, which can be obtained by determining the sampling frequency and required acquisition time. The neglecting of the attenuation coefficient of medium of active path in the calculation of this paper represents a drawback.

In 2009, Leib et al. [17] proposed a pulse-based compact UWB radar used for medical diagnostic, focusing on system architecture, correlation receiver, and time delay adjustment; it is also used for detecting heart beat and respiratory movement. This study supposed that the heart beat and respiratory rates have been fixed, while these ranges have discontinuity. This discontinuity caused a high attenuation in the transmitted signal, causing inaccurate readings.

In 2010, Lazaro et al. [18] estimated vital signs monitor that uses an impulse radio UWB radar; the analytical design has been developed for performing the spectral analysis according to the harmonic and intermodulation addressing for respiration and heart signal, with its simulation and proposed harmonic filter. Finally, the results have been introduced to illustrate the accuracy of the technique for heart rate and respiration detection. The flaw of this study is that the detection is difficult if the frequencies of first breath harmonics and heart signal are being closely.

In 2010, Thiel et al. [19] introduced a UWB sensor to improve the magnetic resonance imaging (MRI) especially for cardiovascular and cancer diagnostic. Their study attempted to prove the benefit of motion tracking for high-resolution brain imaging and navigation used with cardiac MRI, and also served to support electrocardiograph (ECG) analyzing. The proposed device worked only with high or ultra-high MRI field and did not benefit with the low-field MRI.

In 2010, Elmissaoui et al. [20] introduced an imaging radar for human tissue by analyzing the return echoes from the body layers. Their research aimed to find the time of arrival (TOA) and electromagnetic propagation direction (Θ) that basically depends on the characteristic properties of human tissue (layers). The study depended on the reflection echoes to form an image, where the echoes from the deep layers are very weak and difficult to detect because of the attenuation (drawback).

In 2011, Jalilvand et al. [21] examined a UWB system to detect a hemorrhagic stroke in a 3D simulated head model, that estimated four layers model and proving that UWB technique is suited for stroke detection, comparing with other medical imaging devices like MRI, CT scan and mammography. This manner introduced an inaccurate stroke image with low resolution which is considered as main drawback.

In 2012, Urdaneta and Wahid [22] studied a UWB imaging to detect the bone cancer. The main feature of this study is the use of monopole antenna in the frequency range 1–10 GHz based on image reconstruction technique. This measures the change between the dielectric properties of bone tissues and tumor by determining the reflection coefficient, and with a certain algorithm, but still with inaccurate size detection and with long calculation time.

In 2016, Ali et al. [23] designed a noninvasive UWB system for reliable glucose concentration level measurement in human blood depending on artificial intelligence, without taking blood sample, by using two UWB micro-strip antennas with signal acquisition and data processing, where the system works in an artificial neural network manner. The system sends the UWB wave in the central frequency of 4.7 GHz from one side and receives it from the other side, and then applying artificial neural network on the received signal, the drawback of this system is the inaccurate readings and the system needs a long training time.

In 2016, Seguin et al. [24] evaluated a UWB transmission signal to detect heart volume changes with frequency range 1.5–4 GHz, depending on the attenuation changes between blood and other tissues, and using TOA determination for each path of UWB, the drawback of this research is its use of only semi-dynamic heart model in the thorax area, also the propagation, and reflection of other layer.

In 2016, Mackenberg et al. [25] attempted to establish a UWB system that detects the vascular pressure based on the detection of vascular dilation inside an inhomogeneous tissue. The researchers found a correlation relationship between the major peaks of signal in time domain and the amplitude of supposed values. These peaks represent the reflection from the boundary layers of proposed phantom depending on the relative and propagation times. The drawback of this study is that all experiments were applied for homogenous tissue and with one layer (silicon), while the real study must be with multilayer. These layers increased the number of reflected signals due to increasing the complexity of measurements.

In 2017, Wang et al. [26] used impulse-radio (IR) UWB sensor for accurately measuring the chest compression depth. This study uses many trails, which are then compared between them. This study did not respect the human body permittivity which has an effect on the calculation of time of arrival (TOA), and it requires heavy instruments in developmental prototype.

In 2018, Alhawari [27] presented lung tumor detection using UWB (microwave imaging approach) that uses microstrip antenna with frequency range 3–4 GHz with best distance of 10 mm far from the thorax. This introduces a radar with the ability of tumor detection of 4 mm diameter in size, which is very accurate comparing with another chest imaging device like ultrasound and X-ray, and with low cost. The used technique is based on the comparison of the dielectric properties of normal tissue and cancerous tissue, where the cancerous tissue has higher dielectric properties than the normal tissue. The drawback of this research is that the blood and muscle tissues have high dielectric properties compared to lung tissue. Also, the experiment needs accurate adjustment for successive examination and the cancerous tissue has different dielectric properties from one case to another.

In 2018, Der et al. [28] produced a UWB radar based on microwave technique with oblique projection and Rao detectors combination to detect breast tumor. This technique is used for reducing the cumbersome clutter and detecting the existence of a tumor, where the tumor region denotes the maximum power; the drawback of this manner is that it is not quite distinguishable in the low signal-to-noise ratio (SNR) case.

In 2018, Selvaraj et al. [29] proposed an UWB antenna and microwave scattering for early breast tissue tumor detection and localization and finding the depth of tumor with frequency range 2.4–4.7 GHz. The study is based on the reflected signal which is received at a microstrip antenna. The signal passed through tumor is attenuated more than the surround normal tissue. This proposed antenna is inactive when the tissue under test is a heterogeneous material.

In 2018, Aziz et al. [30] introduced a graphene-based conductor UWB patch antenna to detect brain tumor. This antenna is operated at a frequency range of 3.15–9.15 GHz depending on the changes between high-reflection coefficient of normal tissue and low-reflection coefficient of cancerous tissue, as well as based on varying the ground patch width with respect to the value of SAR caused radiation. It is also noted that the antenna length will have an effect on the bandwidth produced, where the length be 7 mm and the antenna is operated as narrowband.

In 2018, Wang [31] proposed an electromagnetic imaging for brain stroke detection based on the changes in the electrical impedance of human tissue in frequency 1–4 GHz with the use of scattered signal to produce a microwave image (MI). The main drawback is that when the stroke is near the skull, it causes an increase in the skull-induced distortion and the system is complicated multi-input multi-output (MIMO).

In 2018, Lee et al. [32] used an IR-UWB radar for monitoring heart rate and rhythms (noncontact). Also, their result's reliability and validity with ordinary ECG are compared. The percentage of mean error is 2.3% vs. 0.2% of normal ECG, which means the UWB radar is inaccurate comparing with the normal ECG. The researchers used MATLAB program to synchronize and store radar readings with normal ECG readings.

In 2018, Shen [33] used a IR-UWB radar to measure the respiration and heart beat rate. The study is based on the autocorrelation, that is, applying fast Fourier transform (FFT) to obtain the respiration rate easily, while reapplying the autocorrelation method after dividing the received signal to the sets of bins and removing one block is the resulting the heartbeat rate signal detection, where the pleural periodical movement caused by the periodicity is displayed as a drawback.

In 2019, Shyu et al. [34] proposed a UWB radar sensor to detect breathing and heart rate. They used First valley peak of the energy function of intrinsic mode functions (FVPIEF) based two-layer ensemble empirical mode decomposition (EEMD). This technique serves the feature time index to detect the frequency of heart beat rate equal to about 1 Hz, which is affected by separating the heart rate from the large breath rate (respiratory). The drawback of this technique is that the breathing movement always masks off the heart beat rate and it is still hidden in the large harmonics and noise.

In 2019, Alghanimi et al. [35] proposed noninvasive blood glucose measurement depending on the relatively changing in the blood dielectric properties by using one ultra-wideband transceiver with a frequency range of 5GHZ and calculating the reflection coefficient through the comparison between the transmitted and reflected signals; the drawback of this study is that there are many factors that can have an effect on the readings of that device like body temperature, gender, blood group, and others.

In 2020, Alghanimi et al. [36] proposed an ultra-wideband radar for angiography by using two different types of antennas. The first antenna is placed around the human body and the other is inserted into the blood vessel in front of the guidewire of catheterizing angiography. The distance between antennas will be measured by calculating the time of arrival and propagation direction, which will be depending on the ultra-wideband frequency, shape, and other specifications. This distance between the antennas includes the human tissue with its different layers, where each layer has certain dielectric properties enabling us to recognize the tissue type. The drawback of this study is the difficulty of manufacturing a small UWB antenna that can be inserted into the human vessels.

3. UWB medical radar for angiography

Many commonly used medical imaging devices for cardiovascular imaging has been found such as X-ray angiography, cardiac MRI, cardiac CT, and cardiac ultrasound (echo). These devices have some limitations such as radiographic exposure, high cost, high complexity, long time requirement, limited resolution, or other medical preventions. These limitations motivate us to begin this research and attempt to find a new technique that avoids the ionizing radiation as well as minimize the cost and complexity. The ultra-wideband technology with its distinctive specifications is the suitable technique.

3.1 Medical UWB radar methodology

The medical cardiovascular imaging UWB radar is a new approach for medical multi-static radar which depends on the use of two different transceivers. The first transceiver is a horn type that has been built around the body of angiography (encloses the human body), while the other is a micro-strip which has been inserted into the human blood vessel with the angiography guidewire. The wave pulses of medical radar traveled through the human tissues and then arrived at the receiver. According to the high dielectric properties, differences among the human tissues, and the blood, the radar can recognize the blood in the vessels, where the high percentage of water tissues (like blood) has high dielectric properties in comparison with other tissues. Also, the reflection coefficients' amplitude and time of peaks have been affected by the depth of blood vessels, thus making the ultra-wideband radar so favorable for cardiovascular imaging. The formation of the medical image is required for the finding of the distance between the two antennas, the propagation direction (Θ), and the time of arrival (TOA), which will be different from one tissue to another, depending on the dielectric properties (permittivity) of the tissue. The wave pulses passing through tissues with high dielectric properties (like blood) will be faster than when it passes through tissues with low dielectric properties. Also, the layers (tissues) have been recognized, depending on the finding of reflection pulses at the time of arrival and propagation direction.

Figure 4 illustrates the possible scenarios for wave propagation (transmission) through the body layers; these scenarios are explained in the conclusion. The new medical radar has been worked instead of the X-ray angiography. To minimize the biological side effects of X-ray RF, which provides accurate real-time imaging which is necessary at catheterization angiography operation for clear imaging, the new medical radar is designed according to the standard model IEEE 802.15.3a channel parameters that consider a fifth derivative Gaussian pulse ultra-wideband with a frequency center of 5 GHz. These pulses have been passed through a hypothetical medium (which represent the human tissues); this medium has been represented by the additive white Gaussian noise (AWGN) and medium gain with a certain delay, where the delay and propagation direction in real cases will depend on the tissues that are passed through. Finally, the received signal will be compared with the transmitted signal by using a cross-correlator. **Figure 5** illustrates this simulation, which represents the MATLAB simulation for the new medical radar with all proposed components.

3.2 Medical radar equations

UWB waves transmit through the body tissue layers (mediums) under electromagnetic wave propagation laws, where the velocity of waves are different from one to another layer according to the dielectric properties (permittivity) of this medium [34, 37]:

$$\mathbf{Vi} = \frac{c}{\sqrt{\varepsilon i}} \tag{4}$$

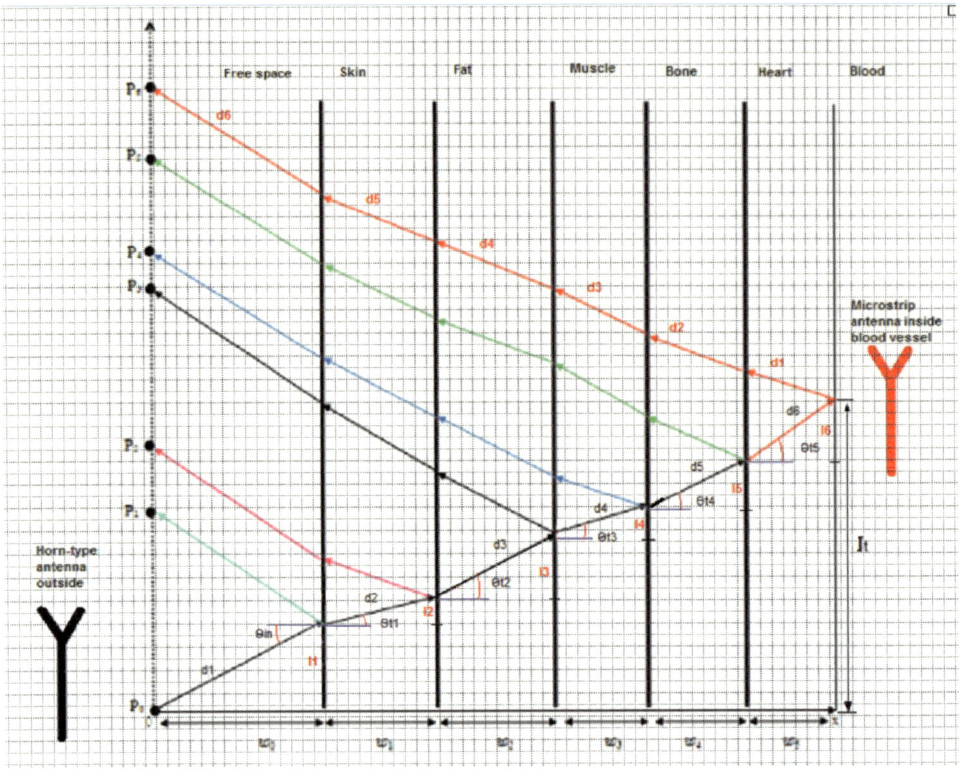

Figure 4.
Distance between two antennas with the including layers.

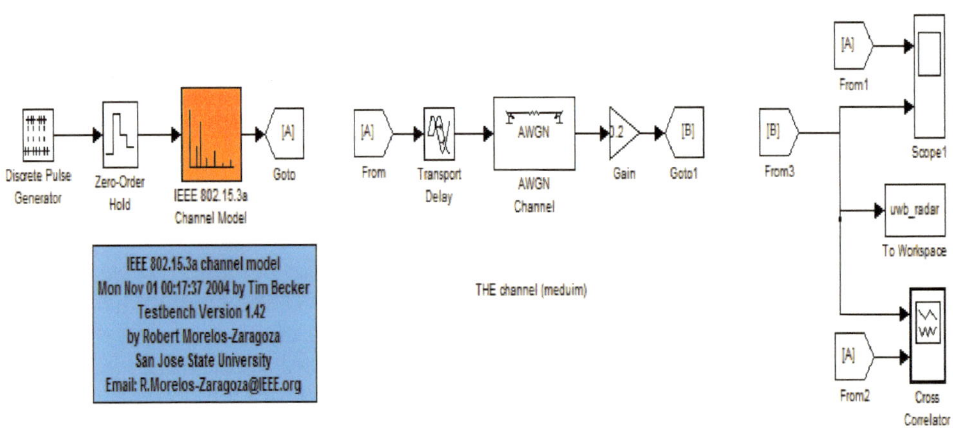

Figure 5.
MATLAB simulation for the proposed UWB medical radar.

where V_i is the velocity in layer i, c is the velocity in free space, and ε_i is the permittivity of the medium. The transmitted angle between two layers will be different from medium to another depending on the intrinsic impedance of the two mediums η_o and η_i and as in **Figure 3** and in [20]:

$$\boldsymbol{\Theta}\mathbf{ti} - 1 = sin^{-1}\left(\frac{\eta o}{\eta i}\sin\left(\Theta in\right)\right) \tag{5}$$

where Θ_{in} is the incident angle which must be greater than the critical angle [37]:

$$\boldsymbol{\Theta c} = sin^{-1}\left(\sqrt{\frac{\varepsilon i + 1}{\varepsilon i}}\right) \tag{6}$$

Here, Θ_c is founded only if the wave transmits from a denser to a less dense layer. And to find the one-way distance between the two transceivers of our radar, we need to find the one-way distance of each layer individually and the distance for the first layer (d_1), second layer (d_2), and any layer (d_i) as the following [20]:

$$\mathbf{d1} = \frac{l1}{\sin(\Theta in)} \tag{7}$$

$$\mathbf{d_2} = d_1 + \frac{l2}{\sin(\Theta t1)} \tag{8}$$

$$\mathbf{d_i} = d_{i-1} + \frac{li}{\sin(\Theta t2)} \tag{9}$$

where l_1, l_2, and l_i are obtained from the following equations:

$$\boldsymbol{l_1} = w_0 \ \tan(\Theta_{in}) \tag{10}$$

$$\boldsymbol{l_2} = w_1 \ \tan(\Theta_{t1}) \tag{11}$$

$$\boldsymbol{l_i} = w_i \ \tan(\Theta_{ti-1}) \tag{12}$$

where ω is the frequency of the wave. The vertical offsets between the two antennas (It) can be obtained by the equation:

$$\boldsymbol{l_t} = \boldsymbol{l_1} + \boldsymbol{l_2} + \boldsymbol{l_3} \ldots. + \boldsymbol{l_i} \tag{13}$$

Also, to find the time $\boldsymbol{t_1}$, $\boldsymbol{t_2}$, and $\boldsymbol{t_i}$ for each layer:

$$\boldsymbol{t_1} = \frac{\boldsymbol{d1}}{\boldsymbol{c}} \tag{14}$$

$$\boldsymbol{t_2} = \boldsymbol{t_1} + \frac{\boldsymbol{d_2} - \boldsymbol{d_1}}{\boldsymbol{v_1}} \tag{15}$$

$$\boldsymbol{t_i} = \boldsymbol{t_{i-1}} + \frac{\boldsymbol{d_i} - \boldsymbol{d_{i-1}}}{\boldsymbol{v_{i-1}}} \tag{16}$$

where Vi is the velocity of wave in the medium. Also, the most important law in our calculations is the intrinsic impedance η for each layer (medium) [38]:

$$\eta = \frac{\sqrt{\mu/\varepsilon}}{\left[1 + \left(\frac{\sigma}{w\varepsilon}\right)^2\right]^{\frac{1}{4}}} \tag{17}$$

where μ is the permeability, ε is the permittivity, σ is the conductivity, and ω is the frequency of wave. Considering the human body tissues as lossy mediums, μ = μo μr, ε = εo εr, $\sigma \neq 0$, where μo is the permeability of free space, μ_r is the relative permeability, ε_o is the permittivity of free space, ε_r is the relative permittivity, and the dielectric properties of free space are: permittivity $\varepsilon_o = 8.854 \times 10^{-12}$ (F/m), permeability $\mu_o = 4\pi \times 10^{-7}$ (H/m), and conductivity σ = 0 [38]. Finally, there are other parameters that can be obtained from the intrinsic impedance, and these parameters are reflection (Γ) and transmission (Γ) coefficients between mediums [34, 39]:

$$\Gamma_{1/2} = \frac{\eta_2 - \eta_1}{\eta_2 + \eta_1} \tag{18}$$

$$T_{1/2} = \frac{2\eta_2}{\eta_2 + \eta_1} \tag{19}$$

Finally, the amplitude of transmitted wave (E_x) will decrease (attenuate) exponentially and can be obtained from the equation:

$$E_x = e^{ax} \tag{20}$$

where x is the crossing distance and a is the attenuation coefficient, while the equations of this wave after incidents at the boundary between the two mediums with different dielectric properties will be:

$$E_t = T.E_i \tag{21}$$

$$E_r = \Gamma.E_i \tag{22}$$

where E_i is the incident wave, E_t is the transmitted wave, E_r is the reflected wave, T is the transmission coefficient, and Γ is the reflection coefficient.

The dielectric properties of human body tissue are estimated by Gabriel [40, 41], so the thicknesses of the tissues (layers) in any region of the human body are represented in [42]. The equations above can be applied on the human body layers depending on the characteristic properties of each tissue which are dependent on the transmitted wave frequency.

3.3 Results and discussion

The above equations have been applied on the human body tissues based on the characteristic properties of the tissues which are dependent on the frequency of transmitted wave.

3.3.1 Intrinsic impedance and transmission angle calculation

The transmission angle between the tissues and the intrinsic impedance of human tissues is calculated as shown in **Table 1**, using a frequency center of 5 GHz, the incident angle of $\pi/4$, and based on the permittivity and conductivity (dielectric properties) of all tissues. Here, the intrinsic impedance has a directional relationship with the transmitted wave frequency, while the transmission angle has an inverse relationship with the frequency of the transmitted wave, and also it is based on the intrinsic impedance and the incident angle.

The results in **Table 1** are essential for time, distance, and speed calculation, which are essentially for the tissue recognition needed for image reconstruction, as mentioned in Section 3.2.

3.3.2 Distance and time calculations

The velocity, time, and the distance between the two transceivers (one-way distance) have been calculated by using the characteristic properties of human

Tissue type	Conductivity σ (S/m)	Permittivity ε (F/m)	Intrinsic impedance η(Ω)	transmission angle θt
Air	1E-20	1	376.734309	
Skin	3.06	35.774	42.7032786	0.24488786
Fat	0.24	5.0291	138.539578	1.17884328
Muscle	4.04	49.54	36.9706969	0.21196925
Bone	0.96	16.05	72.5312559	0.74134990
Heart	4.86	50.27	34.2849767	0.88705125
Blood	5.4	53.95	32.6208079	0.17272680
Lung	3.94	44.859	37.7366007	0.06408701

Table 1.
Dielectric properties of human tissues.

tissues which are listed in **Table 1**. The thickness of these layers (tissues) are taken in the thorax area and ordered as shown in **Figure 4**.

From the results shown in **Table 2**, the variations in the times and speeds for the layers have been observed and are based on the dielectric properties of each layer. These differences will enable the medical image reconstruction depending on the speed of waves in the tissues.

3.3.3 Reflection and transmission coefficient calculation

The reflection coefficient and transmission coefficient with different frequencies are illustrated in **Figures 6** and 7.

From the results which are illustrated in **Figures 6** and 7, note that the reflection and transmission coefficients have a directional relationship with the frequency of a transmitted wave; and from **Figure 6**, note that the blood has the lowest reflection coefficients, which means the ultra-wideband pulses spend shorter time passing through the blood. Also, the transmission coefficient of skin-air is smaller than the transmission coefficient of air-skin, so improving the reflection and transmission coefficients increases the ability of the radar imaging process in any direction either from the inside to the outside transceiver or from the outside to the inside transceiver, as illustrated in **Figure 4**.

The radar can make the two processes together for getting a very clear image, which makes the new radar pass the problems of previous radars, which have been

Tissue type	Thickness	L	Distance	Time	Velocity
Air	50	50	58.76	2E-07	3E+08
Skin	1.3	−0.32	60.1	2.2E-07	5E+07
Fat	9.5	22.98	84.97	4.1E-07	1E+08
Muscle	13.5	−2.91	98.78	7.3E-07	4E+07
Bone	6.6	−6.04	107.7	8.5E-07	7E+07
Heart	5.65	−6.93	117.9	1.1E-06	4E+07
Blood	1.2	0.209	108.9	8.7E-07	4E+07
Lung	5.7	−0.37	123.6	1.2E-06	4E+07

Table 2.
Distance and time between the layers.

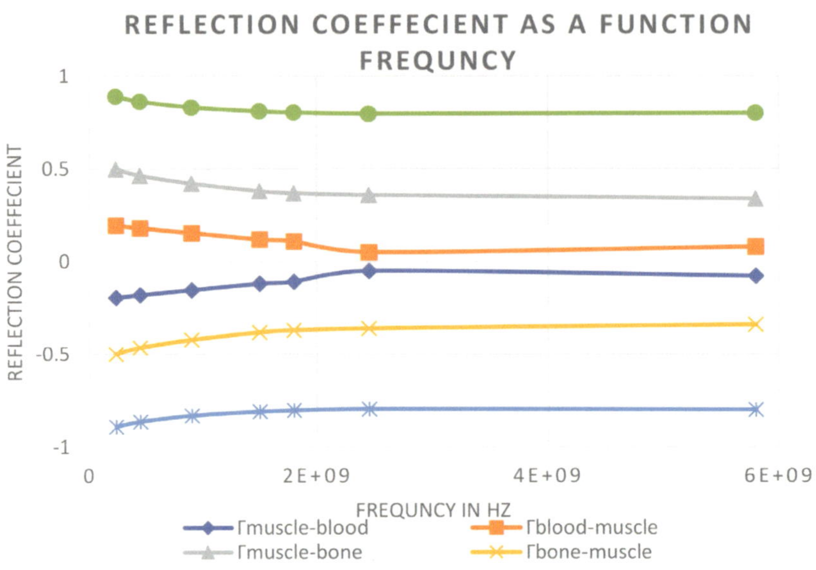

Figure 6.
Relationship between reflection coefficients and frequency.

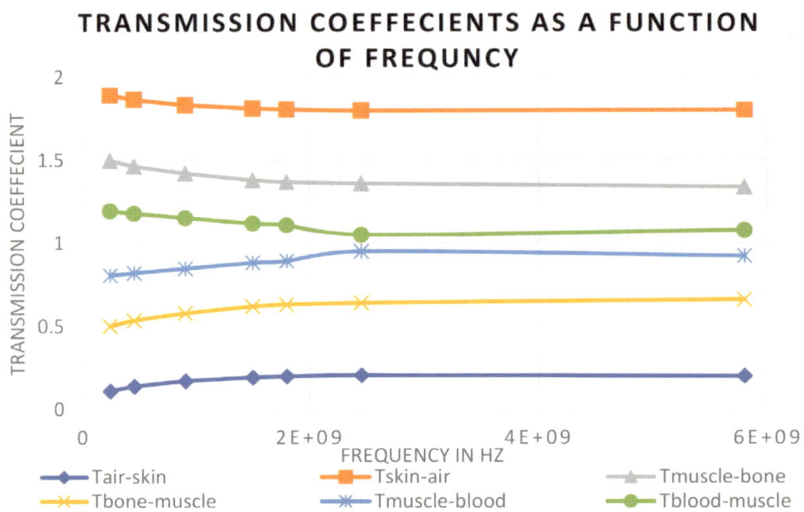

Figure 7.
Relationship between transmission coefficients and frequency.

represented by the power loss in the first layers and the receiving of weak power signals from the other depth layers, also enabling the choosing of the best way that has the lowest reflection coefficient and the highest transmission coefficient due to minimizing of the power dispersion [43].

3.3.4 Experiment results

The first experiment was done by adding the glucose water intravenous nutrient with different concentrations to detect its effects on the dielectric properties of blood (especially on the permittivity), where we will use the glucose concentrations from 70 to 16,000 mg/dL, and **Figure 8** shows the results.

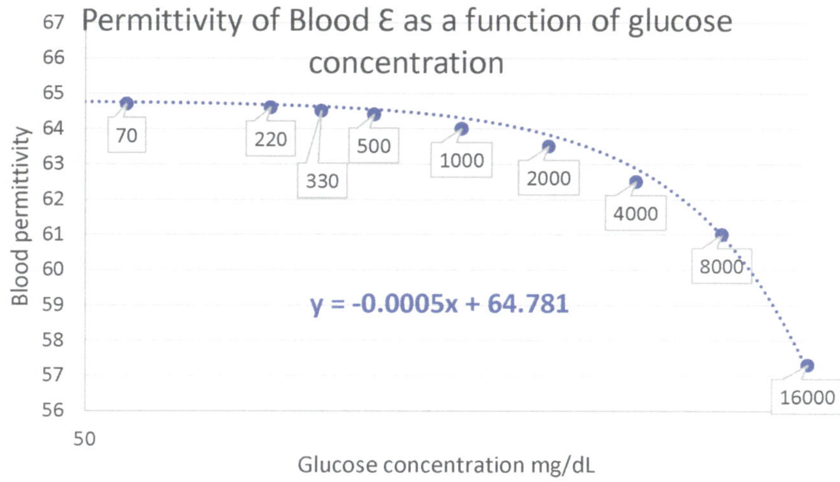

Figure 8.
Permittivity of blood Ɛ as function of glucose concentration.

From the results illustrated in **Figure 8**, the blood permittivity has been decreased gradually when the blood glucose concentration is increased, and the opposite is right, where the blood dielectric properties can be increased by decreasing the blood glucose concentration. Also, these results can be served to design a noninvasive ultra-wideband blood glucose concentration measurement device.

The next experiment has been achieved by adding various normal saline intravenous nutrients' concentration for detecting its effects on the blood dielectric properties (especially the permittivity), where the using of the salinity concentrations from 0 to 2% is applied, and **Figure 9** shows the results. These results have been served to conclude that the blood permittivity will be decreased gradually when the blood salinity percentage, %, is increased, and the opposite is right, where the blood dielectric properties can be increased by decreasing the salinity percentage, %. Also, these results can be served to design a new ultra-wideband device used for detecting the blood salinity percentage noninvasively or for blood pressure measurements.

The last experiment has been achieved by adding the anticoagulant material (citrate or EDTA) to the blood and then measuring its effects on the dielectric properties (permittivity), as illustrated in **Table 3**. From the results above, the adding of anticoagulant to the blood will cause an increase in the blood permittivity.

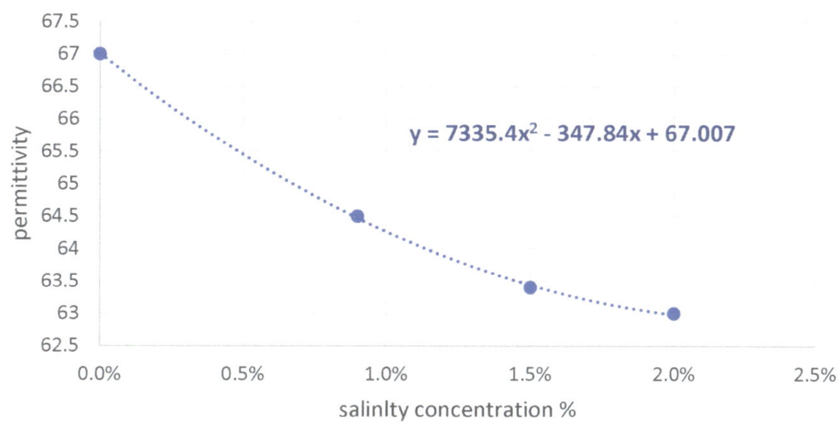

Figure 9.
Permittivity of blood Ɛ as function of salinity concentration.

Pure blood	Blood with citrate	Blood with EDTA
58	62	63.5

Table 3.
Anticoagulant effect.

Note that all experiments have been practiced in the 5 GHz frequency center, with its fitting to the FCC mask, at accepted SAR level, at 37°C temperature, and with fresh blood, and with fresh blood (+O blood group). Also, in the real case, the experiment has some limitations in the adding of these materials to the patients' blood (effected by the patient's condition and his disease background).

4. Ultra-wideband glucometer

It is an ultra-wideband device used for measuring the glucose concentration in the blood noninvasively. The glucose concentration has been found relatively changing within the blood dielectric properties, according to the percentage of glucose that contain. Here, a decrease in the blood dielectric properties (permittivity) has been observed when the glucose percentage was increased (inverse relationship). This decrease was appeared clearly when the frequency range is increased and remained almost constant when it passed the 5GHz as shown as in **Figure 8**. Although, other factors may have an effect on the dielectric properties like temperature, gender, clotting rate, and blood density. This method can be applied accurately by using one ultra-wideband transceiver that is attached to the superficial blood vessel, and then compares the sent and the received signals which have been detected by the same transceiver to calculate the reflection coefficient. The proposed device will be depending on the reflected waves from the blood for measuring the glucose concentration, where a Vivaldi UWB antenna has been attached to the superficial blood vessels to detect the reflection coefficient. The reflection coefficient (Γ) can be obtained from the comparison between the amplitude of transmitted and received (reflected) waves [44]:

$$E_r = \Gamma . E_i \tag{23}$$

where E_i is the incident wave and E_r is the reflected wave. Also, the reflection coefficient has proportional relationship with the blood permittivity (εr) as illustrated below:

$$\Gamma = \frac{\sqrt{\varepsilon_{r1}} - \sqrt{\varepsilon_{r2}}}{\sqrt{\varepsilon_{r1}} + \sqrt{\varepsilon_{r2}}} \tag{24}$$

Finally, the blood glucose concentration has been measured with high percentage of accuracy, which is related to the changes in the blood permittivity (dielectric properties).

5. Conclusions

The finding of the results and hypothesis in Section 3 have concluded a selective direction ultra-wideband imaging radar with multi-static used for angiography with many features; the most important features for this radar are minimizing the power dispersion to the half, which is causing decrease in the attenuation of the signal at the depth of human tissues, and minimizing the time of arrival (TOA), where the ultra-wideband pulses have been received on the other side where there is no need to pass through human tissue again (one-way). The new radar has been designed to cancel the position calculations of return points, and it just needs to calculate the position of the received antenna on the other side by calculating the total offset (It) which is shown in **Figure 4**. As well as the ability of the new radar, for imaging in both directions (selective direction), makes it avoid the old problems of previous radars with power losses and image collapsing. Also, this new radar can be developed to work with other internal imaging devices like endoscopes and with the alike principle of work and similar design in addition to the study of the ability of improving the radar imaging by injecting the patient's body with a certain substance that manipulates the blood dielectric properties. The finding of the transmission coefficient and reflection coefficient among the multilayer tissue enables the new radar for choosing the best way to the imaging, which has been determined depending on the dielectric properties of tissues under exam. If the ultra-wideband pulses are transmitted from a layer with higher dielectric properties to the next layer with lower dielectric properties, then the reflection coefficient has a negative amplitude and the transmission coefficient has a high positive amplitude according to Eqs. (18) and (19), which will be due to most of the ultra-wideband pulses passing the boundary between the layers and arriving to the next layer, and the little percentage of pulses will be reflected from the boundary and will return in the opposite direction of propagation, and the opposite is right. The experiment results in Section 3.3.4 can introduce additional features that can be used for improving the ultra-wideband imaging through adjusting of the dielectric properties of blood by controlling the reflection and transmission coefficients in accord with the radar requirements. And as mentioned in the above paragraph, if the radar sends pulses from the outer transceiver into the inner one (one-way image), then the transmission coefficient must be improved, while if the radar sends pulses from the outer to another transceiver that is also outside the body (two-way image), then the reflection coefficient must be improved. Finally, the blood dielectric properties can help us to find the glucose concentration noninvasively by using one ultra-wideband transceiver (antenna), depending on the comparison between the transmitted and reflected pulses. The UWB antenna will be attached to the superficial blood vessel to avoid the noise and power attenuation from another human tissue to increase the accuracy of readings (where, the using of two antennas to send and receive the pulses from the other side of the hand will result in the wave being passed through many layers, and the power of the signal will be absorbed, which will cause

Relationship	Notes
Glucose concentration $\alpha \frac{1}{\varepsilon}$	
Glucose concentration α $T_{1/2}$	By considering the blood is 2nd medium
Glucose concentration α $T_{1/2}$	By considering the blood is 2nd medium
Anticoagulant material α $\mathcal{E}$	
Anticoagulant material $\alpha \frac{1}{\Gamma_{1/2}}$	By considering the blood is 2nd medium
Anticoagulant material $\alpha \frac{1}{T_{1/2}}$	By consider the blood is 2nd medium
Blood temperature α $\mathcal{E}$	
Hemoglobin percentage α $\mathcal{E}$	

Table 4.
A summary of relationships.

inaccurate measurement). Taking into the account the other parameters can have effects on the dielectric properties, which are mentioned above in Section 1.3. The permittivity has a clear reverse proportional relationship with blood glucose concentration, which can be used to determine the glucose concentration in blood in high accuracy in comparison with the previous studies. **Table 4** will mention a summary of relationships that serve our conclusions:

Author details

Abdulhameed Habeeb Alghanimi
Ministry of Health, Baghdad, Iraq

*Address all correspondence to: abdorck@yahoo.com

References

[1] Federal Communications Commission. Revision of Part 15 of the Commission's Rules Regarding Ultra-Wideband Transmission Systems, First Report and Order. 2010. pp. 02-17

[2] Ahlbom A et al. Guidelines for limiting exposure to time-varying electric, magnetic, and electromagnetic fields (up to 300 GHz). Health Physics. 1998;**74**(4):494-521

[3] Popa A. An optimization of Gaussian UWB pulses. 10th International Conference on Development and Application Systems, Suceava, Romania, May 27-29, 2010; 50

[4] Xie H et al. A varying pulse width 5 th-derivative Gaussian pulse generator for UWB transceivers in CMOS. In: 2008 IEEE Radio and Wireless Symposium. IEEE; 2008. pp. 171-174

[5] Cavagnaro M, Pisa S, Pittella E. Safety aspects of human exposure to ultra wideband radar fields. In: International Symposium on Electromagnetic Compatibility—EMC EUROPE, 17–21 September. 2012. pp. 149-153

[6] Zhang Y, Zhong L, Tan S, Xu C. Dielectric properties of red blood cell suspensions based on broadband dielectric spectrum. In: 2010 IEEE International Conference on Solid Dielectrics. Potsdam, Germany: IEEE. ICSD; 2010. pp. 1-4. DOI: 10.1109/ICSD.2010.5567195

[7] Salahuddin S, Farrugia L, Sammut CV, O'Halloran M, Porter E. Dielectric properties of fresh human blood. In: Proceedings of 2017 19th International Conference on Electromagnetics in Advanced Applications. Ireland, Galway: IEEE. ICEAA; 2017. pp. 356-359. DOI: 10.1109/ICEAA.2017.8065249

[8] Rauf A. A dielectric study on human blood and plasma. International Journal of Environmental Science and Technology [Online]. 2013;**2**(6): 1396-1400. Available from: http://www.ijset.net/journal/216.pdf

[9] Salahuddin S, Halloran MO, Porter E, Farrugia L, Bonello J, Sammut CV. Effects of standard coagulant agents on the dielectric properties of fresh human blood. IEEE Transactions on Dielectrics and Electrical Insulation. 2017;**24**(5): 3283-3289. DOI: 10.1109/TDEI.2017.006582

[10] Brodie G, Jacob MV, Farrell P. 6 Techniques for measuring dielectric properties. In: Microwave and Radio-Frequency Technologies in Agriculture. IEEE; 2015. pp. 52-77

[11] Mekhannikov AI, Myl'nikov AV, Maslennikova LP. Calibration of a coaxial antenna-probe for microwave dielectric measurements. Measurement Techniques. 2007;**50**(4):425-428. DOI: 10.1007/s11018-007-0087-2

[12] Cavagnaro M, Pisa S, Pittella E. Safety aspects of human exposure to ultra wideband radar fields. In: International Symposium on Electromagnetic Compatibility—EMC EUROPE. IEEE; 2012. pp. 1-5

[13] Cavagnaro M, Pisa S, Pittella E. Safety aspects of people exposed to ultra wideband radar fields. International Journal of Antennas and Propagation. 2013;**2013**:1-8

[14] Staderini EM. UWB radars in medicine. IEEE Aerospace and Electronic Systems Magazine. 2002; **17**(1):13-18

[15] Paulson CN, Chang JT, Romero CE, Watson J, Pearce FJ, Levin N. Ultra-wideband radar methods and techniques of medical sensing and

imaging. Smart Medical and Biomedical Sensor Technology III. 2005;**6007**: 60070L

[16] Staderini EM, Varotto G. Optimization criteria in the design of medical UWB radars in compliance with the regulatory masks. In: 2007 IEEE Biomedical Circuits and Systems Conference. 2007. pp. 53-58

[17] Leib M, Schmitt E, Gronau A, Dederer J, Schleicher B, Schumacher H. A compact ultra-wideband radar for medical applications. Frequenz. 2009; **63**:1-2

[18] In P. Analysis of vital signs monitoring using an IR-UWB radar. Progress In Electromagnetics Research. 2010;**100**:265-284

[19] Thiel F, Kosch O, Seifert F. Ultra-wideband sensors for improved magnetic resonance imaging, cardiovascular monitoring and tumour diagnostics. Sensors. 2010;**10**(12): 10778-10802. DOI: 10.3390/s101210778

[20] Elmissaoui T, Soudani N, Bouallegue R. New radar system in medicine. In: European Signal Processing Conference. IEEE; 2010. pp. 1640-1644

[21] Jalilvand M, Li X, Zwick T, Wiesbeck W, Pancera E. Hemorrhagic stroke detection via UWB medical imaging. In: Proceedings of the 5th European Conference on Antennas and Propagation (EUCAP). 2011. pp. 2911-2914

[22] Urdaneta M, Wahid P. A study of UWB imaging for bone cancer detection. In: Proceedings of IEEE International Conference on Ultra-Wideband. IEEE; 2012. pp. 197-201. DOI: 10.1109/ICUWB.2012.6340479

[23] Ali MS, Khatun S, Kamarudin LM, Shoumy NJ, Islam M. Non-invasive ultra-wide band system for reliable blood glucose level detection. International Journal of Applied Engineering Research. 2016;**11**(14):8373-8376

[24] Seguin M, Bourqui J, Fear E, Okoniewski M. Monitoring the heart with ultra-wideband microwave signals: Evaluation with a semi-dynamic heart model. Biomedical Physics & Engineering Express. 2016;**2**(3):1-10. DOI: 10.1088/2057-1976/2/3/035011

[25] Mackenberg M, Rackebrandt K, Bollmeyer C. Model Reflection and transmission of ultra-wideband pulses for detection of vascular pressure variation and spatial resolution within soft tissues. Biomedical Physics & Engineering Express. 2016;**2**(2):065003

[26] Yu BG, Oh JH, Kim Y, Kim TW. Accurate measurement of chest compression depth using impulse-radio ultra-wideband sensor on a mattress. PLOS One. 2017;**12**(8):1-8

[27] Alhawari ARH. Lung tumour detection using ultra-wideband microwave imaging approach. Journal of Fundamental and Applied Sciences. 2018;**10**(2):222-234

[28] Der Fang L, Fang WH, Yu LH, Chen YT. Breast tumor detection in the microwave imaging with oblique projection and rao detectors. In: Proceedings of 4th IEEE International Conference on Applied Systems Innovation. Taiwan: IEEE. ICASI; 2018. pp. 239–242. DOI: 10.1109/ICASI.2018.8394577

[29] Selvaraj V, Baskaran D, Rao PH, Srinivasan P. Breast tissue tumor analysis using wideband antenna and microwave scattering breast tissue tumor analysis using wideband antenna and microwave. IETE Journal of Research. 2018;**01**:1-11. DOI: 10.1080/03772063.2018.1531067

[30] Aziz MAI, Rana MM, Islam MA, Inum R. Effective modeling of GBC

based ultra-wideband patch antenna for brain tumor detection. In: International Conferences on Computer, Communication Chemical, Materials and Electronic Enginerring. IC4ME2, Bangladesh: IEEE; 2018. pp. 1-4. DOI: 10.1109/IC4ME2.2018.8465492

[31] Wang L. Electromagnetic sensing and imaging for stroke detection. In: Proceedings of the 2nd International Symposium on Computer Science and Intelligent Control. Sweden: ACM. 2018. pp. 1-6. DOI: 10.1145/3284557.3284734

[32] Lee Y, Park J, Choi Y, Park H, Cho S. A novel non-contact heart rate monitor using impulse-radio ultra-wideband (IR-UWB) radar technology. Scientific Reports. 2018;**8**(1):13053. DOI: 10.1038/s41598-018-31411-8

[33] Shen H et al. Respiration and heartbeat rates measurement based on autocorrelation using IR-UWB Radar. IEEE Transactions on Circuits and Systems II: Express Briefs. 2018;**65**(10): 1470-1474. DOI: 10.1109/TCSII.2018.2860015

[34] Shyu KK, Chiu LJ, Lee PL, Tung TH, Yang SH. Detection of breathing and heart rates in UWB radar sensor data using FVPIEF-based two-layer EEMD. IEEE Sensors Journal. 2019;**19**(2): 774-784. DOI: 10.1109/JSEN.2018.2878607

[35] Alghanimi AH, Fayadh RA. Non-invasive blood glucose measurement depending on the blood dielectric properties by using one ultra-wideband transceiver. International Research Journal of Engineering and Technology. 2019;**06**(07):1494-1498

[36] Alghanimi AH, Fayadh RA. Ultra-wideband radar for angiography. IOP Conference Series: Materials Science and Engineering. 2020;**745**(1). DOI: 10.1088/1757-899X/745/1/012086

[37] Taoufik E, Nabila S, Ridha B. The ultra wide band radar system parameters in medical application. Journal of Electromagnetic Analysis and Applications. 2011;**03**(05):147-154. DOI: 10.4236/jemaa.2011.35024

[38] Sadiku MNO. Elements of Electromagnetics. US: Oxford University Press; 2014

[39] Aardal Ø, Paichard Y, Brovoll S, Berger T. Physical working principles of medical radar. IEEE Transactions on Biomedical Engineering. 2013;**60**(4): 1142-1149

[40] Gabriel S, Lau RW, Gabriel C. The dielectric properties of biological tissues: I. Literature survey. Physics in Medicine and Biology. 1996;**41**:2251-2269

[41] Gabriel S, Lau RW, Gabriel C. The dielectric properties of biological tissues: II. Measurements in the frequency range 10 to 20 GHz. Physics in Medicine and Biology.1996;**41**(11):2251-2269

[42] Christ A, Klingenböck A, Samaras T, Goiceanu C, Kuster N. The dependence of electromagnetic far-field absorption on body tissue composition in the frequency range from 300 MHz to 6 GHz. IEEE Transactions on Microwave Theory and Techniques. 2006;**54**(5):2188-2194. DOI: 10.1109/TMTT.2006.872789

[43] Ketata M, Dhieb M, Chaoui M, Lahiani M, Ghariani H. Electric field attenuation of an ultra wide band wave (UWB) during propagation in the human body. International journal of Sciences and Techniques of Automatic Control & Computer Engineering (IJ-STA). 2010;**4**(1):1188-1197

[44] Pearce FJ, Levin N. Ultra-wideband radar methods and techniques of medical sensing and imaging. Smart Medical and Biomedical Sensor Technology III. 2005;**6007**:60070L

Chapter 5

Review of Ultra-Wide Band in Team Sports

José Pino-Ortega and Markel Rico-González

Abstract

The use of valid, accurate and reliable systems is fundamental to warrant a high-quality data collection and interpretation. In 2015, FIFA created a department of Electronic Performance and Tracking systems, collecting under this name the more used tracking systems in team sport setting: high-definition cameras, Global Positioning Systems, and Local Positioning Systems. To date, LPS systems proved to be valid and accurate in determining the position and estimating distances and speeds. However, it is hypothesized that between LPS, ultra-wide band (UWB) is the most promising technology for the future. Thus, this chapter was aimed to make an update about UWB technology in sport: the FIFA's regulation, manufacturer that provide this technology, the research articles that assessed validity and reliability of UWB technology, and the criteria standard for the use of this technology.

Keywords: electronic performance and tracking systems, local positioning systems, UWB, technology, accuracy

1. Introduction

Since the monitoring of match performance is now considered a fundamental part of contemporary team sport's players development, professional soccer clubs invest significant amount of money to nurture elite players [1]. Overall, the quantification of both internal and external training and match load are useful in a practical context to aid game understanding and decision making in relation to individual and collective physical training content and prescriptions [1, 2].

Today, training load monitoring is made thanks to Electronic Performance and Tracking Systems (EPTS) [3, 4], which are classified into three types based on different technologies: Global Positioning Systems (GPS) or Global Navigation Satellite Systems (GNSS) [5–7], semi-automatic video camera systems (VID) [8], and Local Positioning Systems (LPS) [9]. The fact that GPS/GNSS and LPS were not allowed during official competition, together with VID are non-invasive technologies, were the main reasons to VID has been the most used EPTS before 2014. However, the acceptance of the use radio-frequency based technologies during competitions, some installation difficulties of VID, and the possibility to add additional microelectromechanical sensors (MEMS) makes that radio-frequency technologies have become most common in team sport settings [4]. However, since some scientific validity and reliability studies have compared LPS and GPS/GNSS based EPTS [10–12], and they have shown high precision measures using LPS, the use of LPS seems to grow in the future [4, 13, 14].

Indoor positioning wireless technologies are classified into infrared, radio-frequency (Radio frequency identification (RFID), Wi-Fi, Bluetooth and ultra-wide band [UWB]), and ultrasound systems [13]. Among different types of wireless indoor positioning systems, UWB is a promising technology for indoor positioning and tracking [13] and also for outdoor venues where there is no possibility of the surrounding infrastructure interfering in the results [3, 13, 14]. Therefore, the aim of this review was to make an update about UWB technology in sport.

2. Safe and accuracy certificates of UWB in team sports

International Federation of Amateur Football (FIFA) organized an event in which EPTS of those manufacturers that wish an official assessment of its devices´ security may be evaluated under standards and common conditions. All of these providers´ devices evaluated positively are certified with International Match Standard (IMS) license and published in the FIFA's website [15]. In addition, FIFA offer a second certificate under the name "FIFA Quality" certificate, in which manufacturers show the accuracy of their devices against a *Gold Standard* registration system. **Tables 1** and **2** show those manufacturers that provide UWB technology-based devices with IMS or "FIFA Quality" certificates.

Manufacturer	Brand	Test Institute
REALTRACK SYSTEMS SL	WIMU PRO	Victoria University
STATSports Group LTD	APEX POD	Sports Labs Ltd.
Catapult Sports	VECTOR	Sports Labs Ltd.

Table 1.
UWB manufacturers with IMS certificate.

FIFA Quality certificate			
Manufacturer	**Brand**	**Test Institute**	**Certification period**
Catapult Sports	VECTOR (LPS)	Victoria University	05-FEB-20 23-JAN-22
REALTRACK SYSTEMS SL	WIMU PRO	Victoria University	29-NOV-19 23-JAN-22

Table 2.
UWB manufacturers with FIFA quality certificate.

3. UWB in scientific articles

Based on a recently published systematic review about the validity and reliability of LPS technology [9], and additional one added due to its recently publication [16], it may be summarized that three studies used UWB technology with 6 antennae around the field and, in general, 18 Hz [10, 17–19], and one used UWB technology with 8 anchors and 33 Hz [16]. All of them belong to three different manufacturer: Realtrack Systems [10, 17, 18], KINEXON [20], and Ubisense [19].

3.1 Validity of UWB technology

Realtrack Sytems´ UWB (WIMU PRO™, RealTrack Systems, Almeria, Spain) was tested in indoor context to assess its validity, revealing 5.2 cm (0.97%) and 5.8 cm (94%) of mean absolute error (MAE) of all estimations in x- and y- position, respectively [17] (**Table 3**). The same system, in outdoor field showed a MAE of 9.57 cm in x-axis positioning and 7.15 cm in y-axis positioning [18] (**Table 3**). A third study assess the validity of an UWB during linear, circular and zig-zag drills in soccer training in walking and running intensities [10]. The authors showed a bias (%) of 0.55 to 5.85% for determining distance covered, and, moreover, a bias between −0.56 and 0.67 for determining mean velocity [10]. Additionally, this system has been compared with an GNSS revealing lower MAE than satellite-based system (**Table 3**). Athlete tracking technology is continually improving due to developments in microprocessors, data processing, and software [21]. Hence, Realtrack System have provide a new modified UWB with height antennae and 33 Hz, which has been recently compared against a real-measure [16]. The authors showed that the mean difference (MD) was less than 4 cm and in 95% of the cases was between 1 cm and 7 cm. the magnitude of the differences was expressed as 0.28% with real measures as the reference. %CV was less than 1% in all cases (**Table 3**). Despite the fact that Realtrack System (Almería, Spain) has published most of the article, an alternative brand of UWB (Ubisens Series 7000 compact tag) was also tested for its accuracy [19]. The authors also showed sufficient accuracy to test positions of players independently of the length of the recorded runs (**Table 5**). Summarizing, all manufacturers that provide UWB technology have showed acceptable accuracy levels for monitoring the position of players in team sports settings (**Tables 3–5**).

3.2 Reliability of UWB technology

Four studies [16–18, 20] aimed to assess the reliability of LPS based on UWB technology. Hoppe et al., (2018) assessed the reliability calculating the differences between the KINEXON ONE UWB devices of each positioning system (i.e. the between device reliability) though typical error. They found typical errors between 0.1 (criterion variable of 10 m jogging with jump) and 1.7 (criterion variable of 129.6 m entire circuit). The LPS revealed good reliability for the entire distance covered, walking over 10 m and sprinting with change of direction, sprinting over 30-m, sprinting over 5–20 m and theoretical maximal force and horizontal power [20]. In addition, Hoppe et al., [20] compared the results of GPS and UWB, and despite some contradictorily results, comparisons of reliability between the GPS and LPS was mainly favorable to LPS [20].

Regarding to the other commercial UWB based device from RealTrack Systems was tested for its intra- and inter-unit reliability [17]. These tests assisted with understanding the degree of error and the amount of variation between the units. A Mann–Whitney U test was performed to compare differences in the differently designed routes and between devices (i.e. the variation in data measured in one participant or another). Inter-unit reliability (i.e. the difference in using one device or another) was determined using Hopkins's reliability spreadsheet to calculate the percentage typical error of measurement and the intra-class correlation coefficient (ICC) values. The intra-unit reliability of UWB in mean velocity varied between 0.895 and 0.999 of ICC (95% of confidence interval) and the low and upper (for inter-unit variability) ranged between −0.09 and 0.42%. In the case of distance covered, the typical error of UWB varied between 0.94 and 4.87% and the lower and upper bias was between

Ref.	Article's information	Outcomes	What this Document Add?
Bastida-Castillo et al. [10]	• Aim: accuracy/reliability • Environment: outdoor. • Algorithm: TDOA. • Number of anchors: 6 • Sampling frequency: 18 Hz. • Gold Standard: timing gates and real measurement. • Drill: linear, circular and zig-zag course.	• *Distance covered* = bias: 0.57–5.85%; Test–retest reliability %TEM: 1.19; Inter-unit reliability bias: 0.18. • *Velocity* = *b*ias: 0.09; ICC: 0.979; bias: 0.01.	• In static conditions and over prolonged periods of time UWB is more accurate than GPS. • GPS accuracy was slightly more affected by the speed and type of displacement than UWB technology. • Intra- and inter-unit reliability was acceptable for both systems analyzed.
Bastida-Castillo, et al. [17]	• Aim: accuracy/reliability. • Environment: indoor. • Algorithm: TDOA. • Number of anchors: 6 • Sampling frequency: 18 Hz. • Gold Standard: Fixed reference lines of basketball court. • Drill: (1) static position; (2) perimeter markings of the court; (3) middle line of court; (4) exterior perimeter of the painted lines; (5) circle 6.75 m line.	• MAE of all estimations for the x-position of 5.2 ± 3.1 cm and for the y-position of 5.8 ± 2.3 cm. • Inter-unit reliability and ICC = 0.65 (x coordinate) and 0.85 (y coordinate).	• Position estimations are very precise and acceptable for tactical analyses. • The error of the position estimations does not change significantly across different courses. • The use of different devices does not significantly affect the measurement error.

Ref.	Article's information	Outcomes	What this Document Add?
Bastida-Castillo, et al., [18]	• Aim: accuracy. • Environment: outdoor. • Algorithm: TDOA. • Number of anchors: 6 • Sampling frequency: 18 Hz. • Gold Standard: GIS. • Drill: (1) perimeter of the field; (2) halfway line; (3) centre circle; (4) perimeter of the penalty area; (5) semicircle penalty area; (6) *SSG* (7 vs. 7).	• MAE = 9.57 ± 2.66 cm (x coordinate) and 7.15 ± 2.62 cm (y coordinate). • *SSG:* For tactical variables, differences between UWB and GPS reached 8.31% (ES = 0.11).	• UWB-20 Hz has been recommended as accurate technology for estimating position of players on the pitch, while GPS-10 Hz has substantial limitations. • Significance differences reported in tactical analysis between GPS and UWB that the error of using one system or another can mean a difference of more than 8%. • Test–retest reliability and inter-unit reliability were good for the two systems assessed.
Pino-Ortega et al., [16]	• Aim: accuracy/reliability. • Environment: indoor. • Algorithm: TDOA. • Number of anchors: 8 • Sampling frequency: 33 Hz. • Gold Standard: real measurement. • Drill: over the lines of the court (linear course).	• The MD was less than 4 cm and in 95% of the cases was between 1 cm and 7 cm, the reference being the real measure. The magnitude of the differences was expressed as 0.28% with real measures as the reference. %CV was less than 1% in all cases. • It is remarkable that %CV was less than 1% in all cases, in going, coming back and in total. • Besides, inter-unit, test–retest and inter-subject analysis did not influence the reliability results.	• An eight antennae UWB system can be considered suitable for locomotion and positioning in an indoor environment.

Table 3.
Studies that assess validity or reliability of Realtrack systems´ UWB (Almería, Spain) (adapted from Rico-González et al. [9]).

Ref.	Article's information	Outcomes	What this Document Add?
Hoppe et al., [20]	• Aim: accuracy/reliability. • Environment: indoor. • Brand: 1.0, Munich, Germany. • Algorithm: not defined. • Number of anchors: 12 • Sampling frequency: 18/20 Hz. • Gold Standard: not defined. • Drill: Specific circuits: walking, jogging, and sprinting sections that were performed either in straight-lines or with changes of direction.	• *Distance covered* UWB 18 Hz: TEE: 1.6–8.0%; CV: 1.1–5.1% UWB 20 Hz, TEE: 1.0–6.0%; CV: 0.7–5.0% • *Sprint* UWB 18 Hz, TEE: 4.5–14.3%; CV: 3.1–7.5% UWB 20 Hz, TEE: 2.1–9.2%; CV: 1.6–7.3% • *Relative loss of data sets due to measurement error* UWB 18 Hz = 20.0% UWB 20 Hz = 15.8%	• Overall, 20 Hz UWB had superior validity and reliability than 18 Hz UWB and 10 Hz GPS.

Table 4.
Studies that assess validity or reliability of KINEXON's UWB (Munich, Germany) (adapted from Rico-González et al. [3]).

Ref.	Article's information	Outcomes	What this Document Add?
Leser et al., [19]	• Aim: accuracy. • Environment: indoor. • Brand: Ubisens Series 7000 Compact Tag. • Algorithm: TDOA/AOA. • Number of anchors: 6 • Sampling frequency: 4.17 ± 0.01 Hz per-tag. • Gold Standard: trundle wheel. • Drill: *Runs* in the center of the playing field and at the borders; *Matches* (5 vs. 5 + 1 player (without ball contact) leading a trundle wheel).	• *Runs* = difference with trundle wheel: 8.25 ± 4.07%; 95% LoA: 0.27–16.22%). • *Match* = MD = 3.45 ± 1.99%; 95% limits of agreement = −0.46–7.35%.	• UWB had enough accuracy for time-motion analysis.

Table 5.
Studies that assess validity or reliability of Ubisense's UWB (Munich, Germany) (adapted from Rico-González et al. [3]).

−2.65 and 2.06%. Thus, it was concluded that the UWB was reliable for distance covered and mean velocity [17]. Another study testing inter-unit reliability of UWB of the RealTrack system presented ICC values of 0.65 and 0.88 for x- and y-axis, respectively [17]. In the last published article using a develop device of this provider, Pino-Ortega et al., found remarkable that %CV of a 33 Hz and 8 antennae UWB was less than 1% in all cases, in going, coming back and in total. Besides, inter-unit, test–retest and inter-subject analysis did not influence the reliability results. Therefore, both KINEXON ONE and Realtrack Systems provide a reliable device for measures in sport settings (**Tables 3** and **4**). The characteristics of Realtrack Systems′, KINEXON ONE's and Ubisense's UWB devices have been summarized in **Table 6**.

	REALTRACK Wimu Pro	KINEXON	Ubisense
Sampling rate	< 55 Hz	10–1000 Hz	Not fixed
N° anchors	6–12	6–16	Scalable
GPS integrated	Yes	No	No
Triaxial accelerometer	4 sensors <1000 Hz	200 Hz	—
Triaxial gyroscope	3 sensors <1000 Hz	200 Hz	—
Triaxial magnetometer	160 Hz	20 Hz	—
Battery life	5	6	—
HR data available	Compatible with 3rd party sensors	Compatible with 3rd party sensors	No
Thresholds for each player?	Yes	Yes	Yes
Real-data available	Yes	Yes	Yes
Raw data available	Yes	Yes	Yes
Visualization platform	Software; App; Online	Software; Online	Software; Online
Technology based	ANT+	—	—

Table 6.
Characteristics of devices based on UWB (extracted from Serpiello [22]).

4. Principles for positioning detection

Radio-frequency EPTS are based on quite similar principles of use for positioning detection [13, 14, 21], however, UWB replace satellite navigation networks by a set of antennae installed in a known positioning around the field in which the data are going to be recorded. Thus, UWB system calculate position of devices using: (1) the antennae set (which act as a reference system), and (2) the devices tracked (**Figure 1**). The communication stablished between antennae allows a detection of each device enclosed in a tight-fitty garment commonly located between each player′ scapulae. So, UWB is based on a wireless technology, which establish a communication in the absence of a physical medium [23]. Concretely, the reference system is composed by a set of antennae located around the field in which the measurements are going to be recorded. Though an algorithm (see 4.2. section) (e.g. Time Difference of Arrival (TDOA)), at least three antennae stablished a circumference around themselves, whose radius is defined by the distance between an antenna and the object [13]. It is known that player positioning is in any place of the circumference's perimeter. When at least three antennae stablish their computation, the circumferences perimeters meet in a common place, where the player is (**Figure 1**).

These communication is stablished using electromagnetic waves which carry data [23]. The values of the electromagnetic waves that allow positioning computation are measured over time, and represented by curves, called sinusoids [24]. These curves appear in a certain shape according to their values. Mathematically, these sinusoids are the result of the number of beats or cycles per second (frequency), the power of each frequency component (amplitudes), and the delay or advantage of a signal (phase), which describe the angular displacement of two sinusoidal functions [23, 25]. The key to transmitting the information is through the use of waves with more complex shapes, as a result of a combination of different sinusoids [25]. Depending on the frequency of these waves, indoor positioning wireless technologies are classified into different types (see introduction section).

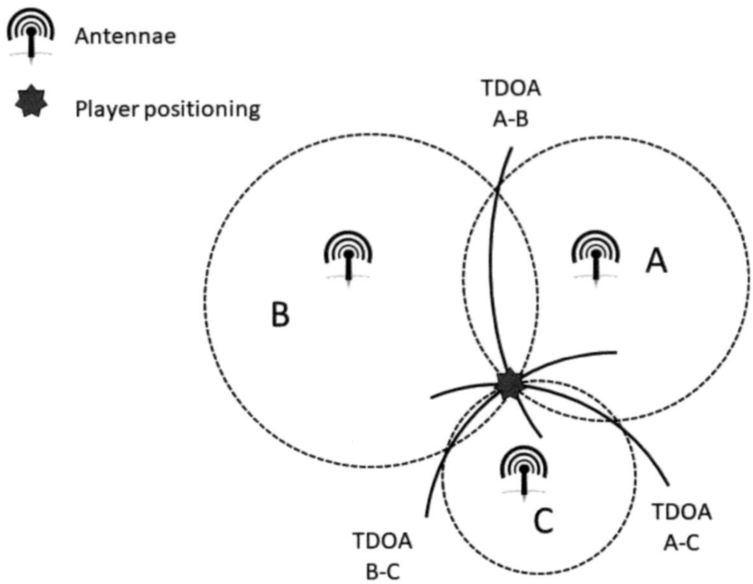

Figure 1.
Positioning using UWB technology.

The distances between antennae (node located in known positioning) and device (held by each player and located in unknown positioning) are computed by UWB positioning algorithms, clustered into different categories: angle of arrival (AOA), received signal strength (RSS); time difference of arrival (TDOA); time of arrival (TOA), and a hybrid algorithm [13]. An understanding of the accuracy, environment, estimation technique, space, and purpose of use of these algorithms is critical because of their differences and the appropriateness of their use in different situations [3, 13]. In brief, despite the fact that AOA algorithm has valid accuracy, AOA and RSS are more suitable than other methods for those systems based on a narrowband signals than with a high UWB bandwidth [3, 13]. Instead, TOA algorithm is suitable for those systems with bandwidth such as UWB. Regarding to the accuracy, small errors in AOA will negatively impact precision when the target object is far away from the base station. However, TOA and TDOA are more accurate relative to other algorithms because of the high time resolution of the UWB signals. In addition, due to hybrid algorithms combine the advantages of all algorithms, it seems to be the most effective solutions for UWB positioning systems [13].

The receiver and transmitter devices that these technologies contain are interconnected to avoid communication with interference from other devices [13]. The communication of the UWB system occupies a very large frequency band, at least 0.5 GHz, as opposed to more traditional radio communications which operate on much smaller frequency bands. On the other hand, since UWB is only allowed to transmit at very low power, its signal emits little noise and can coexist with other services without influencing them (Bastida-Castillo, Gómez-Carmona, De la Cruz-Sánchez, et al., 2019).

5. Limitations and future ways of the use of UWB in team sport

As was analyzed in this chapter, each manufacturer provides a different LPS based on different engineering specifications. Hence, the comparison between LPS provides a wide conclusion due to the comparison between two systems with different standards is difficult. However, since LPS measurement seems to be sensible by several factors such as temperature, humidity gradients, air circulation, pitch dimensions or infrastructure condition, among others [3, 13], the comparison between the outcomes of two studies should be made with caution, even though both of them were performed with the same UWB device. In fact, the same device has resulted in different outcomes in outdoor and indoor context, even in two indoor environments (see **Table 3**). In order to open a research way with the aim of unification all possible information about the use of UWB technology (among others) in the scientific articles′ methodology description, a survey has been published [3] based on the following literature: [5–7, 13, 14, 21, 26–34]. Recently, the information provided in articles has been analyzed based on the survey, and it has been highlighted a need to more detailed descriptions [3]. Accordingly, the items provided by the survey belongs to five quality metrics: (1) system accuracy and precision; (2) coverage and its resolution; (3) latency in making location updates; (4) building's infrastructure impact; and (5) effect of random errors on the system such as errors caused by signal interference and reflection [35]. This fact makes that the comparison between two studies may be unsuitable, at least, while the narrow information is reported. The sampling frequency, computation methods for velocity and acceleration, data exclusion and inclusion criteria, high-intensity bias due to random error, the time at which the data were extracted, technology lock, and data synchronization, and other factors such as the athlete's clothes, the number of reference point, environmental and infrastructure conditions, antennae installation and position, and measurement methods have also been mentioned for the use and description for UWB technology. However, the most of these questions has been addressed in other context such as engineering, and this survey focused in sport settings was based on theoretical framework. The unification of these information will allow a summary of future systematic reviews comparing the outcomes extracted with the same characteristics, and context.

6. Concluding remarks

Theoretically, UWB seems to be the most promising technology for team sports tracking monitoring, however, since it has not been compared against another LPS in team sport setting, it should be considered with caution. In any case, the devices based on UWB technology have shown a high degree of validity for all variables based on positioning (static positioning, time-motion, high speed running and collective tactical behavior). Specifically, Realtrack Systems (6 antennae/18 Hz) = bias (distance covered): 0.55–5.85%, bias (velocity): −0.56 - 0.67, and difference with other EPTS (collective tactical analysis): 8.31%; Realtrack Systems (8 antennae/33 Hz) = bias: 0.28%; KINEXON ONE = TEE: 1.0 ± 6.0%; Ubisense = bias: 8.25 ± 4.07%). Hence, all Realtrack Systems′, KINEXON ONE's and Ubisense systems′ UWB are considered a valid technology for sport settings. Moreover, Realtrack Systems′ and KINEXON ONE's UWB showed to be reliable (KINEXON ONE = TE: 1.7 cm; Realtrack Systems (6 antennae / 18 Hz) = ICC: 0.65 (x-axis) and 0.88 (y-axis); Realtrack Systems (8 antennae / 33 Hz) = %CV: <1%). Therefore, UWB is considered a valid and reliable EPTS in the field of load monitoring of team sports in both indoor and outdoor environments. However, although UWB has usually resulting in greater accuracy than other radio frequency systems at high intensity drills [10], special care should be taken when analyzing load indicators at high speeds or involving different trajectories.

To date, due to the low amount of information reported in the articles′ methodology sections [3], the comparison between outcomes extracted from devices with different characteristics, or in different environment should be made with caution. Therefore, we encourage the authors to explain the methodology about the use of UWB sensors, among others EPTS, based on recently published guideline [3].

Abbreviation list

ANT+	advanced and adaptive network technology
AOA	angle of arrival
EPTS	electronic performance and tracking systems
FIFA	International Federation of Amateur Football
GPS	global positioning systems
GNSS	global navigation satellite systems
Hz	Hertz
ICC	intraclass correlation coefficient
IFAB	International Football Association Board
IMS	International Match Standard
LPS	local positioning systems
MAE	mean absolute error
ME	mean difference
MEMS	microelectromechanical sensors
RFID	radio frequency identification
RSS	received signal strength
TE	typical error
TEE	typical error of estimate
TDOA	time difference of arrival
TOA	time of arrival
UWB	ultra-wide band
VID	semi-automatic video camera systems
%CV	% of coefficient of variation

Author details

José Pino-Ortega[1,2]* and Markel Rico-González[2,3]*

1 Faculty of Sports Sciences, University of Murcia, San Javier, Spain

2 BIOVETMED and SPORTSCI Research Group, University of Murcia, Murcia, Spain

3 Department of Physical Education and Sport, University of the Basque Country, UPV/EHU, Vitoria-Gasteiz, Spain

*Address all correspondence to: josepinoortega@um.es and markeluniv@gmail.com

References

[1] Palucci Vieira LH, Carling C, Barbieri FA, Aquino R, Santiago PRP. Match running performance in young soccer players: A systematic review. Sports Medicine. 2019;**49**:289-318

[2] Rico-González M, Mendez-Villanueva A, Los Arcos A. Training load periodization in soccer with one official match a week: A systematic review. In: An Essential Guide to Sports Performance. New York: NOVA Science publisher; 2020. pp. 123-166

[3] Rico-González M, Los Arcos A, Rojas-Valverde D, Clemente FM, Pino-Ortega J. A survey to assess the quality of the data obtained by radio-frequency technologies and microelectromechanical systems to measure external workload and collective behavior variables in team sports. Sensors. 2020;**16**

[4] Rico-González, M., Pino-Ortega, J., Nakamura, F. Y., Moura, F. A., Rojas-Valverde, D., & Los Arcos, A. (2020). Past, present, and future of the technological tracking methods to assess tactical variables in team sports: A systematic review. *Proceedings of the Institution of Mechanical Engineers, Part P: Journal of Sports Engineering and Technology* 175433712093202. https://doi.org/10.1177/1754337120932023

[5] Castellano J, Casamichana D. Deporte con dispositivos de posicionamiento global (GPS): Aplicaciones y limitaciones. Revista de Psicología del Deporte. 2014;**23**:10

[6] Cummins, C., Orr, R., O'Connor, H., & West, C. (2013). Global Positioning Systems (GPS) and Microtechnology Sensors in Team Sports: A Systematic Review. *Sports Medicine,* 43(10), 1025-1042. https://doi. org/10.1007/s40279-013-0069-2

[7] Hausler, J., Halaki, M., & Orr, R. (2016). Application of Global Positioning System and Microsensor Technology in Competitive Rugby League Match-Play: A Systematic Review and Meta-analysis. *Sports Medicine,* 46(4), 559-588. https://doi.org/10.1007/s40279-015-0440-6

[8] Castellano J, Álvarez-Pastor D, Bradley PS. Evaluation of research using computerised tracking systems (Amisco and Prozone) to analyse physical performance in elite soccer: A systematic review. Sports Medicine. 2014;**44**:701-712

[9] Rico-González, M., Los Arcos, A., Clemente, F. M., Rojas-Valverde, D., & Pino-Ortega, J. (2020). Accuracy and reliability of local positioning Systems for Measuring Sport Movement Patterns in stadium-scale: A systematic review. Applied Sciences 10(17), 5994. https://doi.org/10.3390/ app10175994

[10] Bastida-Castillo, A., Gómez-Carmona, C. D., De la Cruz Sánchez, E., & Pino-Ortega, J. (2018). Accuracy, intra- and inter-unit reliability, and comparison between GPS and UWB-based position-tracking systems used for time–motion analyses in soccer. European Journal of Sport Science, 18(4), 450-457, https://doi.org/10.1080/174613 91.2018.1427796

[11] Linke, D., Link, D., & Lames, M. (2018). Validation of electronic performance and tracking systems EPTS under field conditions. PLoS One, 13(7), e0199519. https://doi.org/10.1371/journal.pone.0199519

[12] Rico-González M, Los Arcos A, Nakamura FY, Gantois P, Pino-Ortega J. A comparison between UWB and GPS devices in the measurement of external load and collective tactical. International Journal of Performance Analysis in Sport. 2020;**20**:10

[13] Alarifi, A., Al-Salman, A., Alsaleh, M., Alnafessah, A., Al-Hadhrami, S.,

Al-Ammar, M., & Al-Khalifa, H. (2016). Ultra wideband indoor positioning technologies: Analysis and recent advances. Sensors, 16(5), 707. https://doi.org/10.3390/s16050707

[14] Leser, R., Baca, A., & Ogris, G. (2011). Local positioning systems in (game) sports. Sensors, 11(10), 9778-9797. https://doi.org/10.3390/s111009778

[15] FIFA. (2018). *About the IMS Standard for Wearable Tracking Devices. Football technology*. https://football-technology.fifa.com/en/media-tiles/about-the-ims-standard-for-wearable-tracking-devices/

[16] Pino-Ortega J, Bastida-Castillo A, Gómez-Carmona C, Rico-González M. Validity and reliability of an eight antennae ultra-wideband local positioning system to measure performance in an indoor environment. *Sports Biomechanics*, 2020. Article in Press.

[17] Bastida-Castillo, A., Gómez-Carmona, C., De la Cruz-Sánchez, E., Reche-Royo, X., Ibáñez, S., & Pino-Ortega, J. (2019). Accuracy and Inter-Unit Reliability of Ultra-Wide-Band Tracking System in Indoor Exercise. *Applied Sciences,* 9(5), 939. https://doi.org/10.3390/app9050939

[18] Bastida-Castillo, A., Gómez-Carmona, C. D., De La Cruz Sánchez, E., & Pino-Ortega, J. (2019). Comparing accuracy between global positioning systems and ultra-wideband-based position tracking systems used for tactical analyses in soccer. European Journal of Sport Science, 19(9), 1157-1165. https://doi.org/10.1080/17461391.2019.1584248

[19] Leser, R., Schleindlhuber, A., Lyons, K., & Baca, A. (2014). Accuracy of an UWB-based position tracking system used for time-motion analyses in game sports. European Journal of Sport Science, 14(7), 635-642. https://doi.org/10.1080/ 17461391.2014.884167

[20] Hoppe, M. W., Baumgart, C., Polglaze, T., & Freiwald, J. (2018). Validity and reliability of GPS and LPS for measuring distances covered and sprint mechanical properties in team sports. PLoS One, 13(2), e0192708. https://doi.org/10.1371/journal.pone.0192708

[21] Malone, J. J., Lovell, R., Varley, M. C., & Coutts, A. J. (2017). Unpacking the Black Box: Applications and Considerations for Using GPS Devices in Sport. *International Journal of Sports Physiology and Performance*, 12(Suppl 2), 18-26. https://doi.org/10.1123/ijspp.2016-0236

[22] Serpiello, F. (2019). *Compare GPS - upper price tier.* Compare Sports Tech. https://www.comparesportstech.com/compare-gps-tracking-systems

[23] Rico-González, M., Los Arcos, A., Nakamura, F. Y., Moura, F. A., & Pino-Ortega, J. (2020). The use of technology and sampling frequency to measure variables of tactical positioning in team sports: A systematic review. Research in Sports Medicine, 28(2), 279-292. https://doi.org/10.1080/15438627.2019.1660879

[24] Jovanov, E., Milenkovic, A., Otto, C., & de Groen, P. C. (2015). A wireless body area network of intelligent motion sensors for computer assisted physical rehabilitation. *J NeuroEngineering Rehabil*, 2(1), 6. https://doi.org/10.1186/1743-0003-2-6

[25] Winter DA. (2009). Biomechanics and Motor Control of Human Movement. 4. Ed. Hoboken, NJ: John Wiley & Sons, Inc., Hoboken, New Jersey; 2009

[26] Adesida, Y., Papi, E., & McGregor, A. H. (2019). Exploring the Role of Wearable Technology in Sport

Kinematics and Kinetics: A Systematic Review. *Sensors,* 19 (7), 1597. https://doi.org/10.3390/s19071597

[27] Aughey, R. J., & Falloon, C. (2010). Real-time versus post-game GPS data in team sports. Journal of Science and Medicine in Sport, 13(3), 348-349. https://doi.org/10.1016/j.jsams.2009.01.006

[28] Chambers, R., Gabbett, T. J., Cole, M. H., & Beard, A. (2015). The Use of Wearable Microsensors to Quantify Sport-Specific Movements. *Sports Med,* 45(7), 1065-1081. https://doi.org/10.1007/s40279-015-0332-9

[29] Chen, K. Y., Janz, K. F., Zhu, W., & Brychta, R. J. (2012). Redefining the Roles of Sensors in Objective Physical Activity Monitoring: *Medicine & Science in Sports & Exercise,* 44: S13–S23. https://doi.org/10.1249/MSS.0b013e3182399bc8

[30] Hennessy, L., & Jeffreys, I. (2018). The current use of GPS, Its Potential, and Limitations in Soccer: *Strength and Conditioning Journal,* 40(3), 83-94. https://doi.org/10.1519/SSC.0000000000000386

[31] Jackson BM, Polglaze T, Dawson B, King T, Peeling P. Comparing global positioning system and global navigation satellite system measures of team-sport movements. International Journal of Sports Physiology and Performance. 2018;**13**(8):1005-1010. DOI: https://doi.org/10.1123/ijspp.2017-0529

[32] Niu X, Li Y, Zhang H, Wang Q, Ban Y. Fast thermal calibration of low-grade inertial sensors and inertial measurement units. Sensors. 2013;**13**(9):12192-12217. DOI: https://doi.org/10.3390/s130912192

[33] Pons, E., García-Calvo, T., Resta, R., Blanco, H., López del Campo, R., Díaz García, J., & Pulido, J. J. (2019). A comparison of a GPS device and a multi-camera video technology during official soccer matches: Agreement between systems. PLoS One, 14(8), e0220729. https://doi.org/10.1371/journal.pone.0220729

[34] Treviño G. Trilateración: Sismos, GPS, rayos y teléfonos celulares, y la XIX Olimpiada de Ciencias de la Tierra. GEOS. 2014;**34**(2):15

[35] Wu H, Marshall A, Yu W. Path planning and following algorithms in an indoor navigation model for visually impaired. In Proceedings of the Second International Conference on Internet Monitoring and Protection, ICIMP. 2007:38-48

Section 4

Ultra-Wideband Antennas and Devices

Chapter 6

Time Domain Performance Evaluation of UWB Antennas

Gopikrishna Madanan and Deepti Das Krishna

Abstract

The performance of printed wideband antennas has to be optimized both in frequency and time domains, to qualify for UWB applications. This is especially true in multi-resonant antenna topologies where the excitation of different modes can change phase centers and radiation patterns with frequency. The study presented in this chapter intends to demonstrate the simulation and experimental design for the time domain characterization of UWB antennas. Modeling the antenna as a linear time-invariant system with transfer function and impulse response, distortion caused to a nanosecond pulse is analyzed. Two planar monopole antenna designs are considered for the comparative study: the SQMA and RMA. SQMA is a traditional CPW-fed monopole design with ground modifications for ultra wide-bandwidth. RMA is a rectangular CPW-fed monopole with an impedance transformer arrangement at the antenna feed. RMA maintains constant impedance over the entire UWB and contributes towards maintaining uniformity in the radiation patterns over the entire frequency band by its design. Transfer function measurements are performed for both the azimuthal and elevation planes and the impulse responses are deduced by performing IFFT. Parameters such as FWHM and ringing are computed from the impulse response for the performance comparison. To evaluate the influence of the antenna geometry on a transmitted/received pulse, the impulse responses are convoluted with a standard UWB pulse. The time-domain distortion for the designs is then compared by computing the Fidelity parameter.

Keywords: ultra wide band, planar monopole antenna, antenna transfer function, antenna impulse response, time domain analysis

1. Introduction

Ultra-wide band systems transmit and receive ultra-short electromagnetic pulses having limited effective radiated power. The system performance is determined primarily by the characteristics of the radiators that have to conform with stringent frequency and time domain requirements in the entire operating band [1]. These requirements are namely, a non-dispersive phase centre; constant radiation and impedance over the frequency range with no excitation of higher order modes [2]. Planar monopole antennas with wide operating bands are most often well matched multi-resonant structures. In [3], Ma et al. have described the time domain performance of a printed dipole antenna employing a tapered slot feed. It is reported that the received pulses are distorted and broadened to more than 1 ns in spite of the antenna being well matched over the entire band. The higher order modes generated tends to shift the antenna phase centers with frequency and can lead to broadening of the radiated pulse. The same authors in [4] reports a much lesser pulse distortion for a tapered slot antenna operating in the 3.1–10.6 GHz UWB band. The present chapter intends to outline the design, simulation and measurement steps followed to characterize UWB antennas in the time domain.

Two monopole antennas with similar design topology is considered for the present analysis: the SQMA with a square radiator and RMA with a rectangular radiator. In both the antennas, the ground plane is optimally designed for a wide impedance bandwidth. The general practice to realize ultra wide-bandwidth in square/rectangular radiating elements is to modify the ground–radiator interface [5–10]. In SQMA, the inherent resonances of a square patch is matched over the wide band by incorporating cuts in the ground plane at specific locations [11]. The designs proposed in [11, 12] have been contemplated to design the RMA. In RMA, the PCB footprint is greatly reduced and radiation patterns in the 3.1–10.6 UWB is stable without much radiation squinting which is otherise observed in wideband antennas at higher frequencies primarily due to the current distribution on the antenna ground plane [13]. Incase of RMA, this aspect is taken into consideration in the design ensuring broadside radiation all through the band of operation. This is further confirmed from the spatio-temporal transfer characteristics and the transient response analysis, making it suitable for UWB applications.

2. Time domain analysis

Antennas intended for UWB systems need to possess superior pulse handling capabilities. Analysis of the transient response of the antenna is performed by direct time domain measurements [14, 15] or by a frequency domain measurement followed by Fourier Transformation [16, 17]. Frequency domain measurements take advantage of the high dynamic range and the standardized calibration of the vector network analyzer and is equally accurate to the direct time domain measurements.

In the latter method, the antenna is considered as a linear time invariant system described by its transfer function (gain and phase) and the associated impulse response [18, 19]. Transfer function and impulse response of UWB antennas are modeled as spatial vectors as the antenna characteristics depend on the signal propagation direction [16, 17]. **Figure 1** shows the model for the analysis which consists of a radio link made up of two antennas in free space under the approximation of far–field and line-of-sight propagation. In this model, the transmitting and receiving antennas are characterized by eliminating channel effects [20].

2.1 Transient reception

The time domain relation between the received voltage pulse $\vec{U}_{Rx}(\omega, r, \theta, \varphi)$ and the incident electric field pulse $\vec{E}_{rad}(\omega, r, \theta, \varphi)$ is [17],

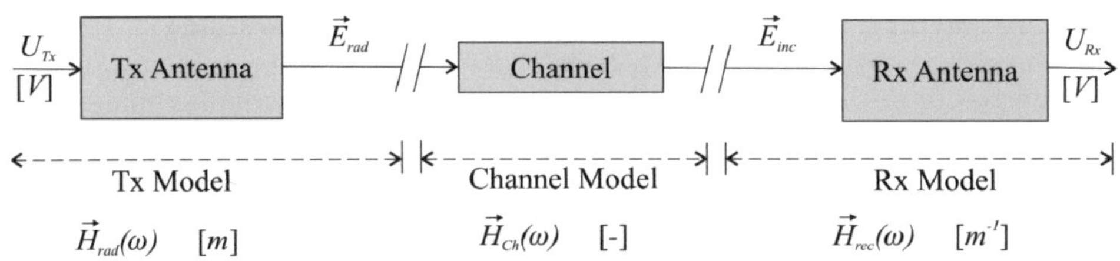

Figure 1.
The UWB channel model.

$$\frac{\vec{U}_{Rx}(\omega, r, \theta, \varphi)}{\sqrt{z_c}} = \vec{h}_{Rx}(\omega, \theta, \varphi) \frac{\vec{E}_{rad}(\omega, r, \theta, \varphi)}{\sqrt{z_0}} \tag{1}$$

In the time domain, the corresponding relation is,

$$\frac{\vec{u}_{Rx}(t, r, \theta, \varphi)}{\sqrt{z_c}} = \vec{h}_{Rx}(t, \theta, \varphi) \otimes \frac{\vec{e}_{rad}(t, r, \theta, \varphi)}{\sqrt{z_0}} \tag{2}$$

z_c and z_0 are the antenna port and free–space characteristic impedance respectively, while $\otimes$ indicates convolution operation. If (θ, φ) represents the elevation and azimuth angles, the receive antenna transfer function $\vec{h}_{Rx}(\omega, \theta, \varphi)$ and its impulse response $\vec{h}_{Rx}(t, \theta, \varphi)$ are considered as a function of the angle of arrival of the received pulse.

An ideal receiving antenna should receive a voltage pulse of the same shape as the one incident on it from any direction. This means that it should have a Dirac–delta impulse response which is also be independent of the angle of arrival. In other words, the antenna transfer function should have a uniform amplitude and a linear phase response (or constant group delay). However, in practice, the receiving antenna transfer function is always band limited.

2.2 Transient radiation

The frequency domain relation between the transmitted electric field pulse and the applied voltage pulse is,

$$\frac{\vec{E}_{rad}(\omega, r, \theta, \varphi)}{\sqrt{z_0}} = \vec{h}_{Tx}(\omega, \theta, \varphi) \frac{e^{-j\omega r/c}}{r} \frac{U_{Tx}(\omega, r, \theta, \varphi)}{\sqrt{z_c}} \tag{3}$$

and,

$$\vec{h}_{Tx}(\omega, \theta, \varphi) = \frac{j\omega}{2\pi c} \vec{h}_{Rx}(\omega, \theta, \varphi) \tag{4}$$

The time domain relation between the transmitted electric field pulse $\vec{e}_{rad}(t, r, \theta, \varphi)$ and the applied voltage pulse $\vec{u}_{Tx}(t, r, \theta, \varphi)$ at the transmitting antenna is,

$$\frac{\vec{e}_{rad}(t, r, \theta, \varphi)}{\sqrt{z_0}} = \frac{1}{r} \partial\left(t - \frac{r}{c}\right) \otimes \left[\frac{1}{2\pi c} \frac{\partial}{\partial t} \vec{h}_{Rx}(t, \theta, \varphi)\right] \otimes \frac{u_{Tx}(t, r, \theta, \varphi)}{\sqrt{z_c}} \tag{5}$$

where,

$$\vec{h}_{Tx}(t, \theta, \varphi) = \frac{1}{2\pi c} \frac{\partial}{\partial t} \vec{h}_{Rx}(t, \theta, \varphi) \tag{6}$$

The convolution with the delta function in Eq. (5) represents the time delay attributed to the finite speed of light, denoted by c. Eq. (6) indicates that the transmit impulse response $\vec{h}_{Tx}(t, \theta, \varphi)$ is a time derivative of the receive impulse response $\vec{h}_{Rx}(t, \theta, \varphi)$. As a result, an ideal antenna (with $\vec{h}_{Rx}(t, \theta, \sigma) = \delta(t)$) will radiate an electric field pulse which would be a first–order time derivative of the input voltage pulse.

2.3 Model of transient transmission

The full input-to-output characteristics, in both frequency and time domains is deduced from 1 and 3 as below,

$$\begin{aligned}\frac{\vec{U}_{Rx}(\omega,r,\theta,\varphi)}{U_{Tx}(\omega,r,\theta,\varphi)} &= \vec{A}_{Tx,Rx}\,\frac{e^{-j\omega r/c}}{r}\,\vec{h}_{Rx}(\omega,\theta,\varphi) \\ &= \sqrt{\frac{2\omega}{c}}\,\vec{A}_{Tx,Rx}\,\frac{c\,e^{-j\omega r/c}}{2r\omega}\,\sqrt{\frac{2\omega}{c}}\,\vec{h}_{Rx}(\omega,\theta,\varphi)\end{aligned} \tag{7}$$

where,

$$\vec{A}_{Tx,Rx} = \frac{j\omega}{2\pi c}\,\vec{h}_{Rx}(\omega,\theta,\varphi)$$

$$S_{21}(\omega) = \frac{\vec{U}_{Rx}(\omega,r,\theta,\varphi)}{U_{Tx}(\omega,r,\theta,\varphi)} = \vec{H}_{Tx}(\omega,\theta,\varphi)\,H_{Ch}(\omega)\,\vec{H}_{Rx}(\omega,\theta,\varphi) \tag{8}$$

where,

$$\vec{H}_{Tx}(\omega,\theta,\varphi) = \sqrt{\frac{2\omega}{c}}\,\vec{A}_{Tx,Rx} \tag{9}$$

$$H_{Ch}(\omega) = \frac{c\,e^{-j\omega r/c}}{2r\omega} \tag{10}$$

$$\vec{H}_{Rx}(\omega,\theta,\varphi) = \sqrt{\frac{2\omega}{c}}\,\vec{h}_{Rx}(\omega,\theta,\varphi) \tag{11}$$

In the time domain, 7 reads,

$$\frac{\vec{u}_{Rx}(t,r,\theta,\varphi)}{u_{Tx}(t,r,\theta,\varphi)} = \vec{h}_{Tx}(t,\theta,\varphi) \otimes h_{Ch}(t) \otimes \vec{h}_{Rx}(t,\theta,\varphi) \tag{12}$$

2.4 Simulation

For the time domain characterization of the antennas, the fourth derivative of the Gaussian pulse given in Eq. (13) and shown in **Figure 2(a)** is chosen as the input pulse, as this pulse conforms to the FCC spectral mask as shown in **Figure 2(b)** when A = 0.333 and T = 0.175 nS.

$$s_i(t) = A\cdot\left[3 - 6\left(\frac{4\pi}{T^2}\right)(t-\tau)^2 + \left(\frac{4\pi}{T^2}\right)^2 (t-\tau)^4\right] e^{-2\pi\left(\frac{t-\tau}{T}\right)^2} \quad \left(\frac{V}{m}\right) \tag{13}$$

For the antennas discussed in the later sections of this chapter, the input voltage $u_{Tx}(t)$ (= $s_i(t)$) is specified in CST Microwave Studio and the radiated pulse $\vec{e}_{rad}(t,r,\theta,\varphi)$ is calculated on a sphere of radius 25 cm. Fourier transforms of these two quantities are then calculated and the transfer function $\vec{H}_{Rx}(\omega,\theta,\varphi)$ is

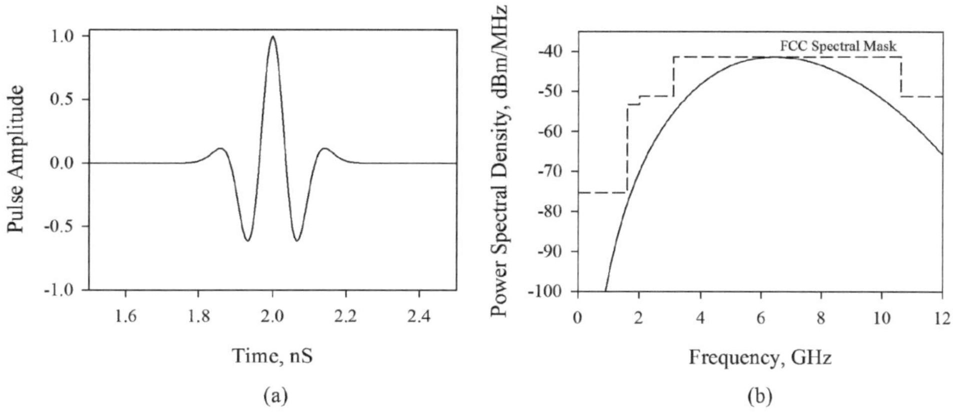

Figure 2.
(a) Input pulse (b) power spectral density.

deduced from Eq. (4). An IFFT operation on this would generate the pulse response $\vec{h}_{Rx}(t, \theta, \varphi)$.

2.5 Measurement

A scattering parameter S_{21} measurement in the frequency domain can be used to deduce the transfer functions of both the transmitting and receiving antennas. Two identical standard horn antennas oriented in the bore sight direction is used, and considering Eq. (7), the corresponding transfer functions are found out from

$$H_{Tx}(\omega, \theta, \varphi) = \sqrt{\frac{j}{2\pi}\left(\frac{\omega}{c}\right)^2 \frac{S_{21}(\omega, \theta, \varphi)}{H_{Ch}(\omega, \theta, \varphi)}} \qquad \left(m^{-1}\right) \tag{14}$$

$$H_{Rx}(\omega, \theta, \varphi) = \sqrt{\frac{2\pi}{j}\left(\frac{c}{\omega}\right)^2 \frac{S_{21}(\omega, \theta, \varphi)}{H_{Ch}(\omega, \theta, \varphi)}} \qquad (m) \tag{15}$$

where the free space transfer function is,

$$H_{Ch}(\omega) = \frac{c}{2d\omega} exp\left(\frac{-j\omega d}{c}\right) \tag{16}$$

Once the reference antenna (Tx Antenna) is characterized, transfer function of the AUT is found for multiple orientations using,

$$H_{AUT}(\omega, \theta, \varphi) = \frac{S_{21}(\omega, \theta, \varphi)}{H_{Tx}(\omega) H_{Ch}(\omega)} \tag{17}$$

2.6 Data processing, windowing

S_{21} is measured with frequency resolution 15.25 MHz and Eqs. (14)–(17) are used to compute the corresponding transfer functions. The data is further appended with zero padding for 0–2 GHz and 12–62.47 GHz range with 4096 points in the pass band. The conjugate of this signal is is then reflected to the negative frequencies to get a double sided spectrum of a real signal which is then transformed to the time domain using IFFT.

Measurements using two identical wide band horns are shown in **Figure 3**. While **Figure 3(a)** represents the measured antenna transfer functions, **Figure 3(b)** shows the real and imaginary parts of the transfer functions. Transmitting and receiving antenna impulse responses are shown in **Figure 3(c)** which clearly indicates that $\vec{h}_{Tx}(t)$ is a derivative of $\vec{h}_{Rx}(t)$.

2.7 Time domain parameters

The time domain evaluation of the antenna effects on the transmitted/received signals is carried out by analyzing the envelope of the analytic response for co–polarization. The real valued antenna's transient response is,

$$h_n(t) = \Re \vec{h}_n(t) \tag{18}$$

The peak output voltage from an incident wave form depends on the peak value $p(\theta, \varphi)$ of the antenna's transient response:

$$p(\theta, \varphi) = max\ t\ \left|\vec{h}_n(t, \theta, \varphi)\right| \tag{19}$$

A measure for the linear distortion of the antenna is the envelope width, which is defined as the full width at half maximum (FWHM) of the magnitude of the transient response envelope.

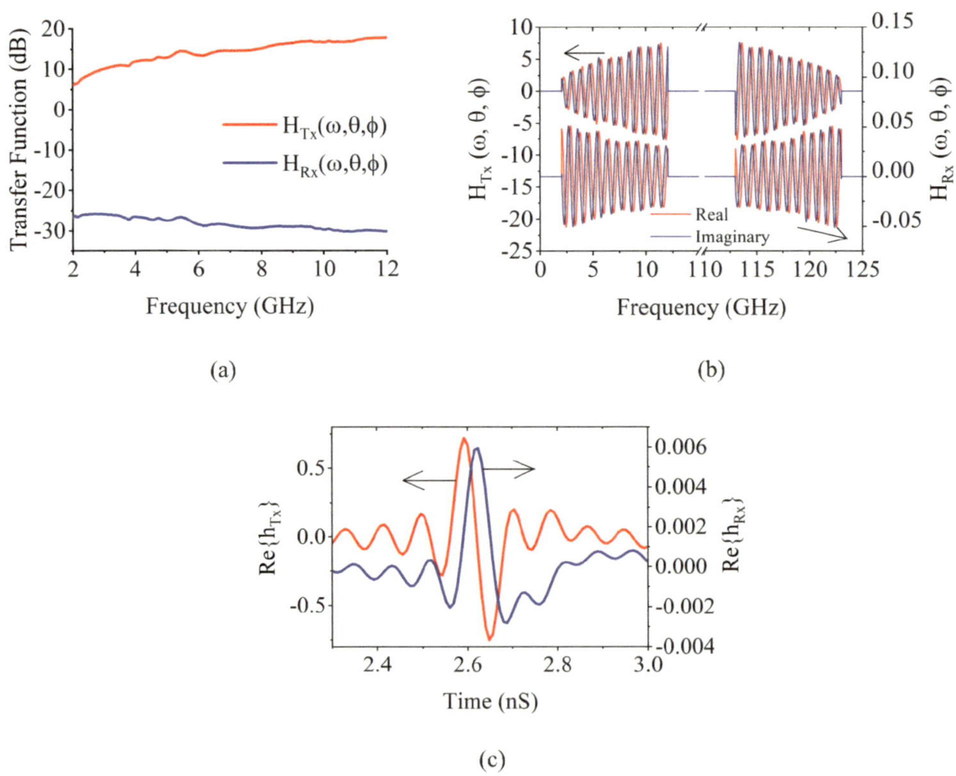

Figure 3.
(a) Transmitting and receiving transfer functions of identical wide band horn antennas (b) transmitting and receiving transfer functions after conjugate reflection (c) impulse response of the transmitting antenna and receiving antenna.

$$FWHM_{0.5}(\theta,\varphi) = t_2\big|_{p/2} - t_1\big|_{p/2,t_1<t_2} \tag{20}$$

The duration of the ringing is defined as the time until the envelope has fallen from the peak value below a fraction α of the main peak.

$$RINGING_\alpha = t_2\big|_{\alpha\cdot p} - t_1\big|_{p,t_1<t_2} \tag{21}$$

The lower bound for α is chosen according to the noise floor of the measurement. In order to compare the ringing of antennas with different gains under the condition of a constant noise floor, the fraction α is chosen to be 0.22 (−13 dB).

2.8 Pulse distortion analysis

The antenna effects on the received signal is evaluated by convoluting the analytic response $\vec{h}_n(t,\theta,\varphi)$ with the input pulse $s_i(t)$.

$$s_o(t) = \vec{h}_n(t,\theta,\varphi) \otimes s_i(t) \tag{22}$$

For UWB systems, receivers are in general based on the pulse energy detection or correlation with the template waveform. Therefore, the pulse distortions can be examined by calculating fidelity factor which is defined as [21],

$$F = max \frac{\int_{-\infty}^{+\infty} s_i(t)\cdot s_o(t-\tau)\,dt}{\sqrt{\int_{-\infty}^{+\infty} |s_i(t)^2|\,dt \cdot \int_{-\infty}^{+\infty} |s_o(t)^2|\,dt}} \tag{23}$$

where τ is the delay parameter varied in a sense to maximize the numerator. The fidelity parameter measurement is performed for different spatial orientations of the test antenna and is evaluated as the maximum of the cross–correlation function. It compares only shapes of the waveforms, not amplitudes.

3. Performance comparison of UWB antennas

To demonstrate the necessity of the time domain analysis of UWB antennas, two topologies shown in **Figure 4** is considered.

3.1 Antenna designs

Figure 4(a) depicts the design of the SQMA. The overall length of the element determines the lower end of the operational band and the upper end by the feed gap '*d*'. It is the ground plane which determines the impedance bandwidth of a monopole antenna. In the present design, the cuts engraved in the ground plane match the multiple resonances, making the antenna operate over the specified ultra wide band. The SQMA has the following dimensions with units of lengths in mm: s = 2.3, g = 0.48, l_patch = 16, l = 14.57, b = 13, d = 0.5, w = 1.4, L = 32, W = 30, ε_r = 4.4, $tan\ (\delta)$ = 0.02, h = 1.6.

The topology of the RMA comprises of a rectangular monopole of area $l_1 \times b_1$ as shown in **Figure 4(b)**. This antenna is evolved from a basic printed monopole design shown in the same Figure. To facilitate ultra wide band performance, an impedance transformer is embedded in the CPW transmission line with increasing slot widths g_1, g_2, g_3 and g_4. The special design of the transformer provide gradual transformation of the 50 Ω line impedance and at the same time, result in a new current path for

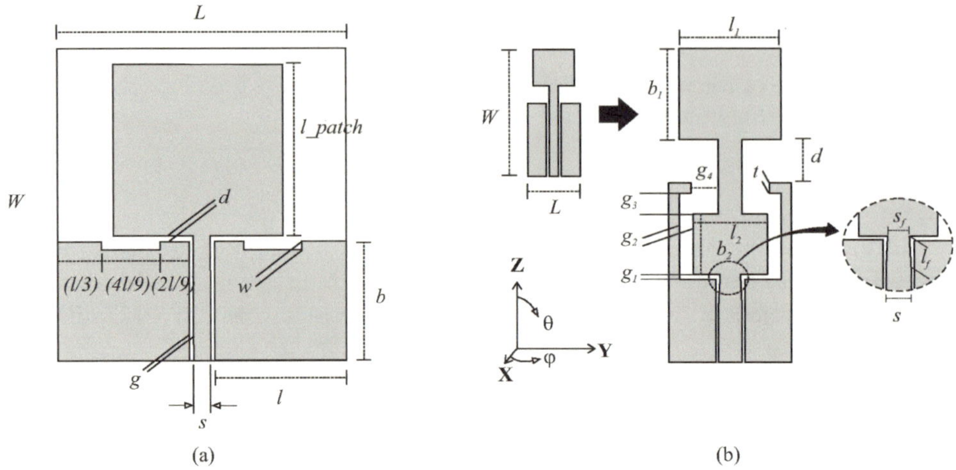

Figure 4.
Design and parameters of (a) SQMA (b) RMA.

radiation. The coupling of the guided waves to the radiator is strongly dependent on parameter d and the resonances in the structure can be matched by suitably choosing d. For the RMA, the geometric parameters are: s = 2.3, g = 0.28, d = 4.25, l_1 = 9, b_1 = 10, l_g = 17.75, l_2 = 6, b_2 = 7.3, g_1 = 0.45, g_2 = 1.35, g_3 = 2, g_4 = 2.7, t = 1, s_f = 2, l_f = 3, L = 30, W = 12, ε_r = 4.4, $tan\ (\delta)$ = 0.02, h = 1.6. Unit of all lengths are in mm.

3.2 Performance evaluation in the frequency domain

Measured and simulated return losses of the two antennas are shown in **Figure 5**. The prominent resonances in the SQMA are at 3.8 GHz, 8 GHz and

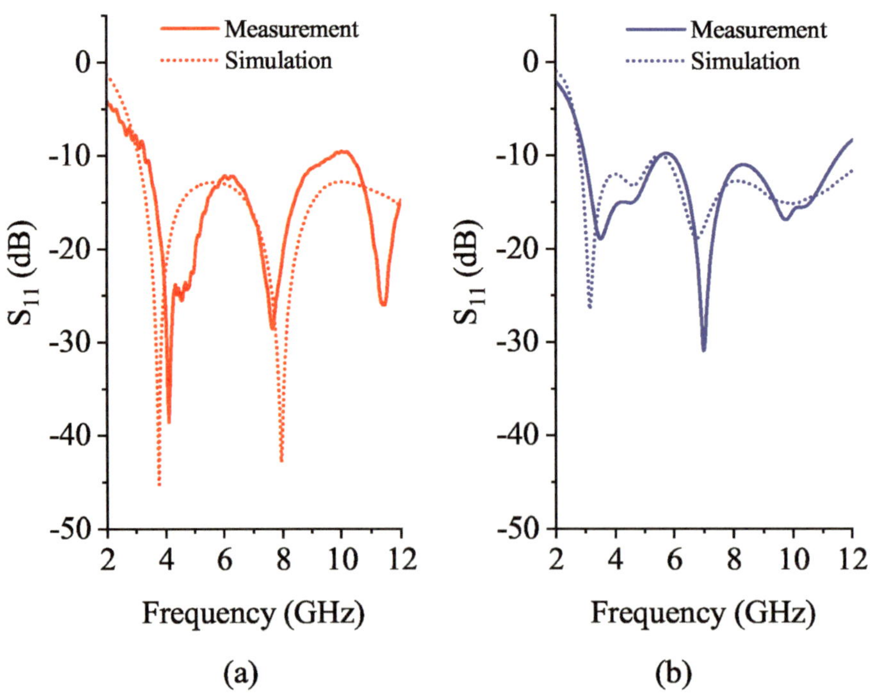

Figure 5.
Measured and simulated S_{11}(dB) of the (a) SQMA and (b) RMA.

13 GHz while that for the RMA are at 3.5 GHz, 5 GHz, 7 GHz and 11 GHz. Both antennas cover the 3.1-10.6 GHz UWB.

3-D radiation patterns of the two antennas obtained from CST are shown in **Figure 6**. In SQMA, the pattern degradation at the higher frequencies were found to be effective from 8 GHz onwards. SQMA also shows pattern squinting as in **Figure 6(a)**. For the RMA, the pattern degradation at the higher frequencies are minimum; radiation is stable and directed towards the bore-sight at all frequencies. The cross–polar level exhibited by both the antennas remain better than −20 dB in the 3.1–10.6 GHz band. Both antennas have linear polarization, oriented in the Z direction. Peak gains of the antennas were measured by gain transfer method. Average value of the peak gain in the 3.1–10.6 GHz UWB is found to be 3.75 dBi for SQMA and 2.6 dBi for the RMA.

3.3 Performance evaluation in the time domain

As an UWB signal is transmitted or received, the pulse shape is altered. This distortion is reflected in the transfer function (in the frequency domain) or impulse response (in the time domain) of the antenna. The antenna transfer function, a complex quantity, would have a constant amplitude and linear phase response while the impulse response would be a delta function, in cases when the antenna has no effect on the pulse transmitted.

3.3.1 Results of simulation in CST

Magnitude of the transfer function $\vec{h}_{Rx}(\omega, \theta, \varphi)$ simulated in CST for the x-y plane of the antennas are shown in **Figure 7(a)** (Using Eq. (3)). It is seen that the intensity plots for $\vec{h}_{Rx}(\omega, \theta, \varphi)$ and $\vec{h}_{Tx}(\omega, \theta, \varphi)$ differ only in their relative

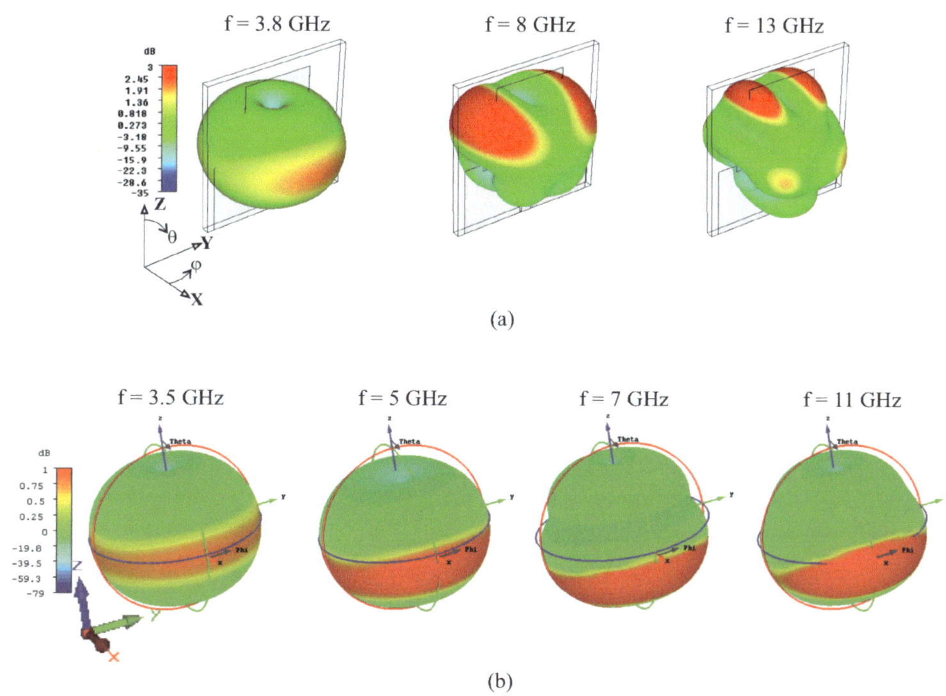

Figure 6.
Simulated radiation patterns at different frequencies for the the antennas (a) SQMA (b) RMA.

amplitudes. Magnitude of the transfer function decay gradually as a function of frequency over most of the azimuth angles (x-y plane). The loss of omni-directionality of the pattern of SQMA at frequencies above 8 GHz is reflected in its transfer function as relatively low intensities at at certain angles. Sharp nulls are seen between 30^0–50^0, 130^0–150^0, 210^0–230^0 and 320^0–330^0. Transfer function in the x-z plane of the SQMA is shown in **Figure 7(b)**. The nulls in the radiation

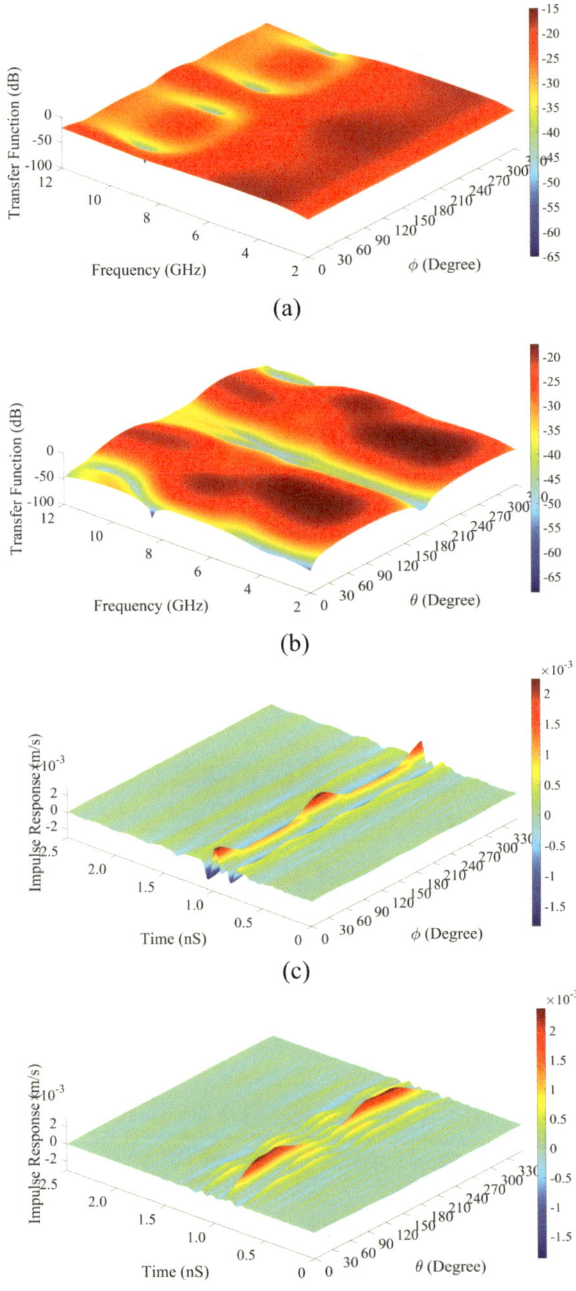

Figure 7.
Computed transfer functions of the SQMA in the (a) x-y (b) x-z planes; computed impulse responses the SQMA in the (c) x-y (d) x-z planes.

pattern at 180^0 can be seen in the transfer functions which show intensity variations.

Impulse response in the x-y plane is well formed with good peak value over almost all the angular regions as shown in **Figure 7(c)**. Amplitude degradation beyond 8 GHz in the SQMA impulse response is reflected between 30^0–120^0

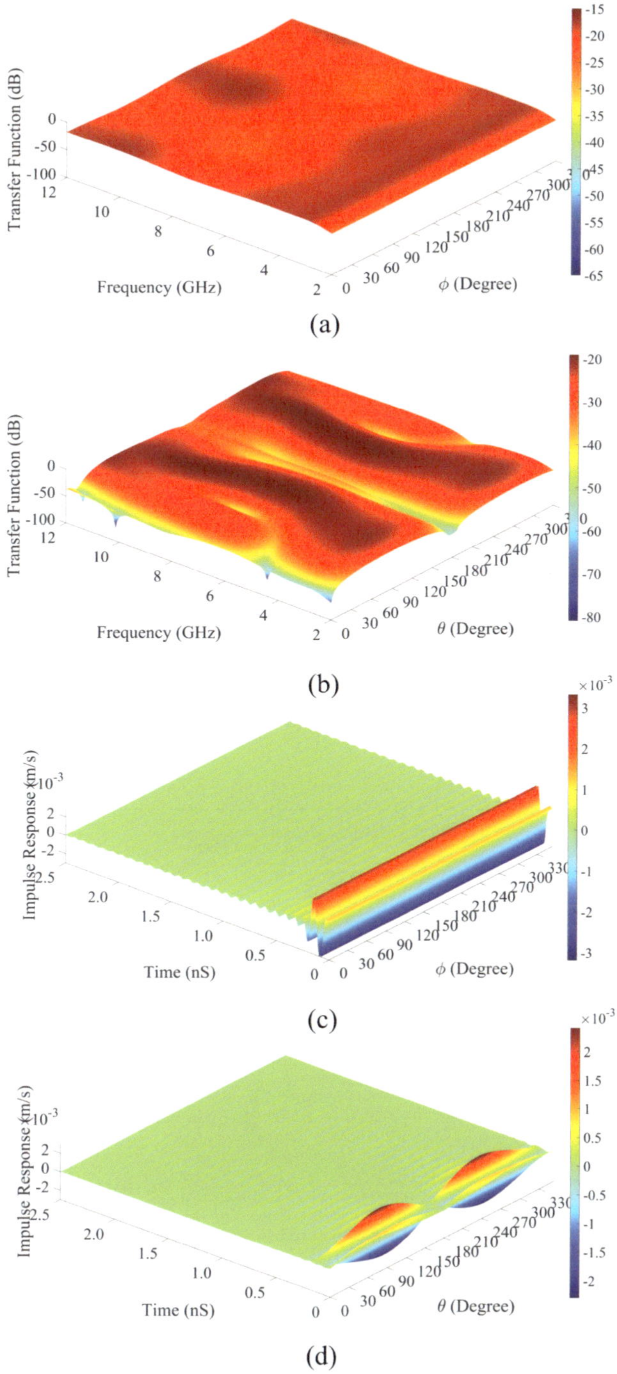

Figure 8.
Computed transfer functions of the RMA in the (a) x-y (b) x-z planes; computed impulse responses the RMA in the (c) x-y (d) x-z planes.

and 200^0–330^0. Impulse response in the x-z plane of the SQMA is shown in **Figure 7(d)**.

Figure 8(a) indicates that the antenna transfer function of the RMA in the x-y plane is fairly constant throughout the entire UWB. Even though computed transfer function in the x-z plane shown in **Figure 8(b)** has variations, it remains within the acceptable limits. The nulls observed in the transfer function in **Figure 8(b)** can be attributed to the monopole pattern of RMA. Impulse responses computed in the x-y and x-z planes for the RMA is shown in **Figure 8(c)** and **8(d)**. In the x-y plane, the impulse response preserves the expected Dirac-delta shape, unlike that in the SQMA. Even though not in the perfect shape, impulse response of the RMA in the x-z plane has more resemblance to a Dirac-delta, when compared to SQMA.

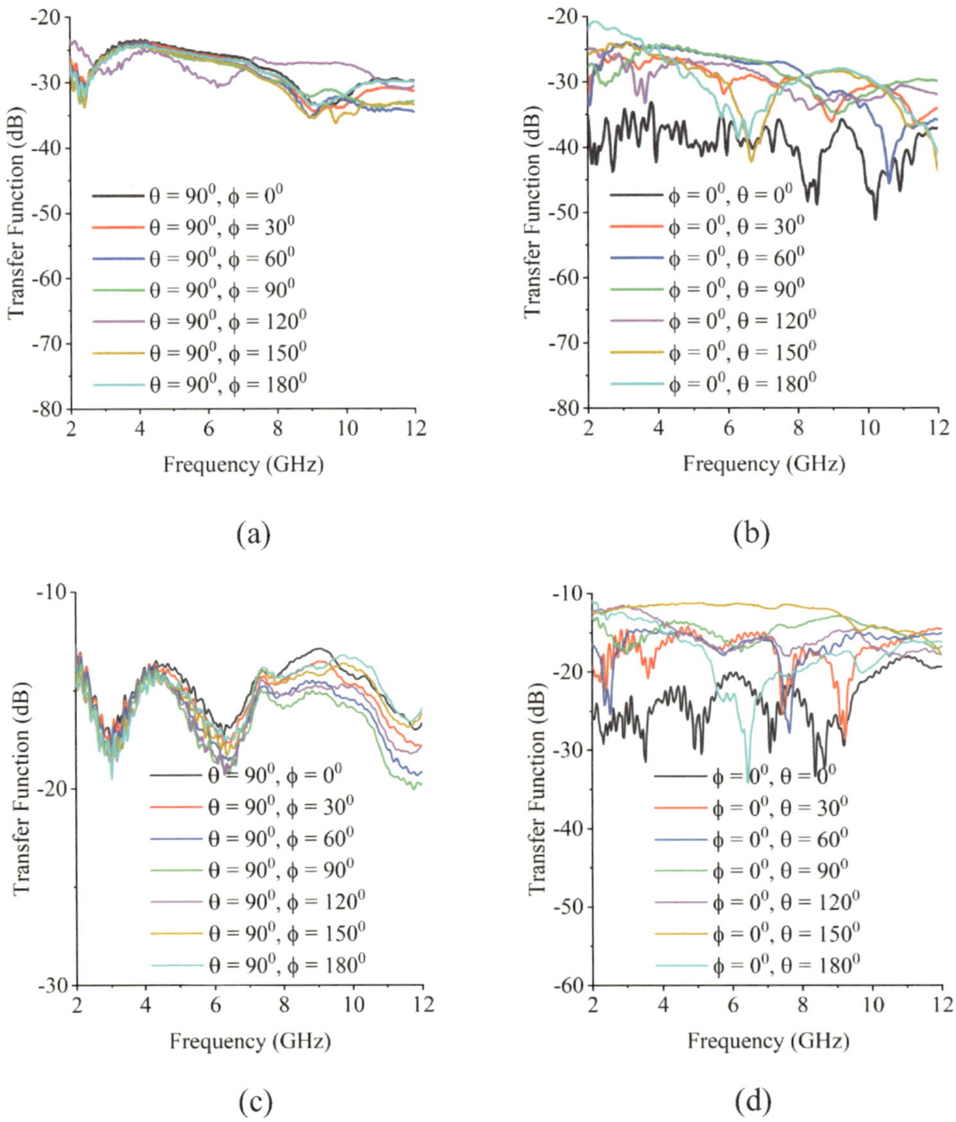

Figure 9.
Measured antenna transfer functions: (a) SQMA, x-y plane (b) SQMA, x-z plane (c) RMA, x-y plane (d) RMA, x-z plane.

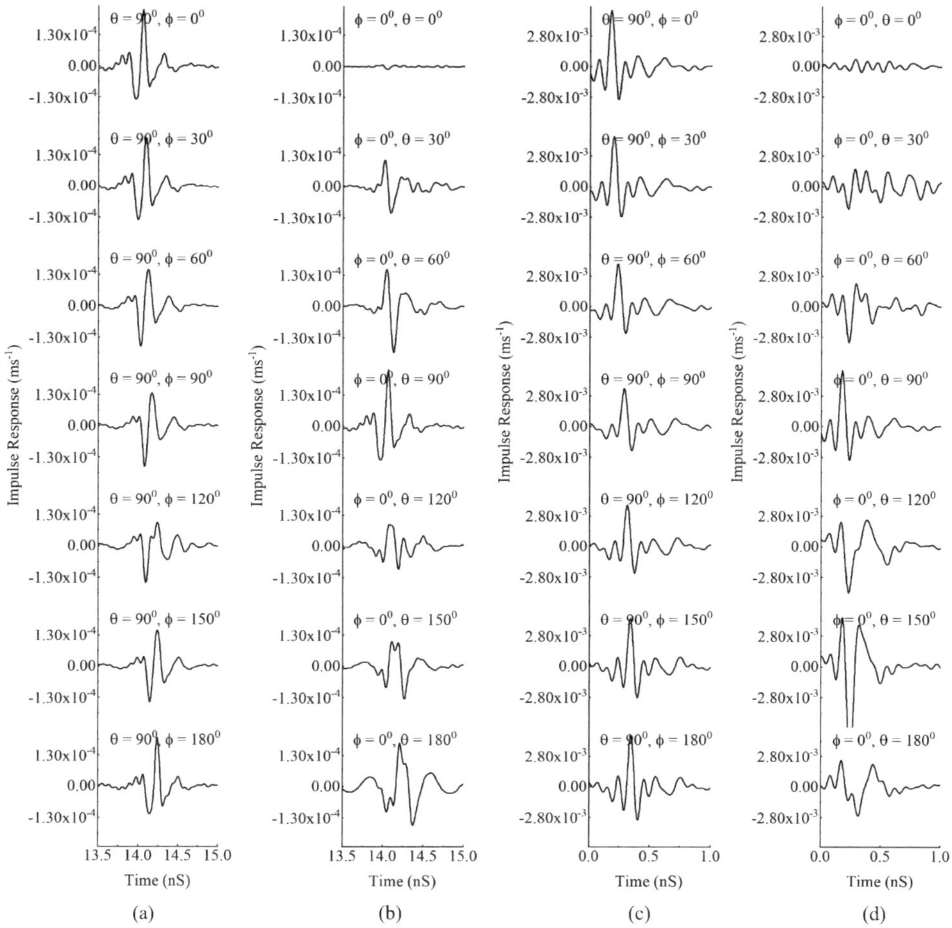

Figure 10.
Measured impulse responses: (a) SQMA, x-y plane (b) SQMA, x-z plane (c) RMA, x-y plane (d) RMA, x-z plane.

3.3.2 Experimental results

For the measurements, the transmitting and receiving antennas were positioned in their far field, at a distance of 25 cm. Source power level in the VNA was set at 10 dB to improve signal to noise ratio in the measured data.

Figure 9(a) and **(b)** indicate the measured antenna transfer functions in the x-y and x-z planes for the SQMA. The measurements seem to follow simulations indicated in Section 3.3.1. **Figure 9(c)** and **(d)** indicate the transfer functions in the x-y and x-z planes for the RMA. While the transfer functions in the x-y planes remain constant with variation within 10 dB, the measured values for the x-z plane shows large variations at particular frequencies.

The **Figure 10(a)–(d)** shows the corresponding impulse responses, obtained by performing an IFFT on the measured transfer functions and they resemble delta functions across all the angles in the x-y plane.

Figures 11 and **12** shows the time domain performance indicators of the antennas for x-y and y-z planes. For the RMA, as the **Figure 11(a)** shows, the FWHM is constant and ringing is minimum at all angles in the x-y plane. The SQMA however shows variations, though within acceptable limits, which could be attributed to its relatively larger size. Computations also futher confirms that fidelity of the received

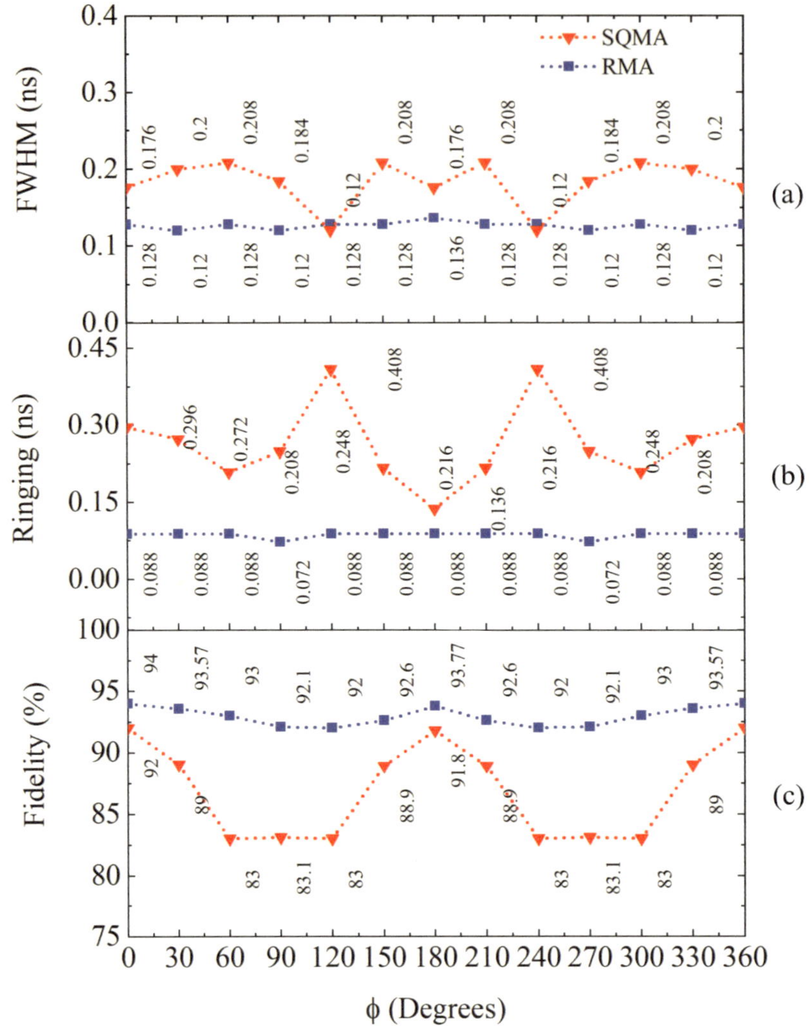

Figure 11.
Computed (a) FWHM (b) ringing and (c) Fidelity measured in the x-y plane.

pulses in the case of RMA is constant and better compared to the SQMA. The performance indicators show variations in the x-z plane in tune with the corresponding radiation patterns. Impulse responses are convoluted with the pulse form given in Eq. (13) to study the effect of the antenna geometry on a transmitted the baseband pulse. Computed fidelity of the received pulses of the RMA is constant and better compared to the SQMA. Variation in Fidelity in the $(\theta = 0^0, \phi = 0^0)$ and $(\theta = 180^0, \phi = 0^0)$ directions could be attributed to the nulls in the radiation patterns at those angles.

4. Conclusion

Printed UWB antennas are multi-resonant structures that operate over a wide frequency range. Even if the radiators designed for UWB exhibit excellent bandwidth, efficiency, etc., traveling phase centers present in the geometry can impair their use in time-domain applications. Changes in the antenna phase centers are reflected in the time domain as a dispersion in the transmitted/received pulses. A

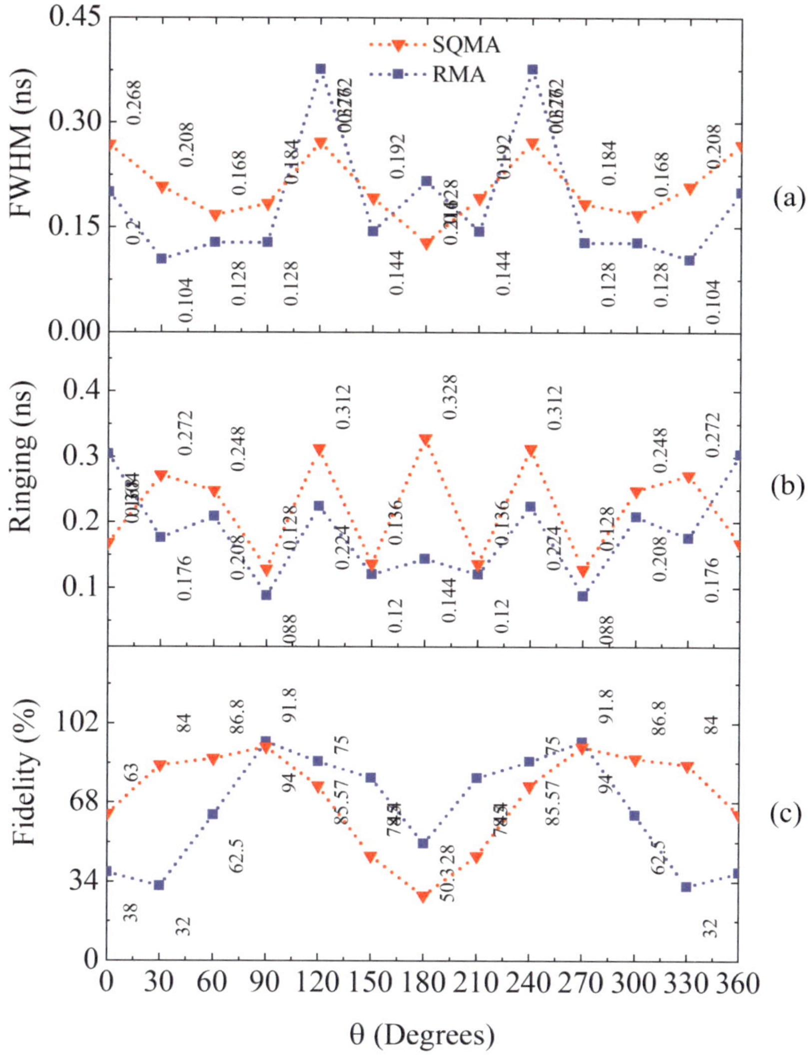

Figure 12.
Computed (a) FWHM (b) ringing and (c) Fidelity measured in the x-z plane.

variation in antenna phase center can be quantified in terms of group delay or antenna transfer function, a curb on the former being 1nS over the entire band and that on the latter is to remain within 10 dB.

In this work, we have demonstrated this point by comparing the time-domain performance of two similar antenna geometries: the SQMA and the RMA. PCB area occupied by the SQMA is approximately 2.7 times that of the RMA and resonances in the geometry are inherent due to the square patch itself. Because of the different modes that are excited in the geometry, its peak radiation points are disoriented in the higher side of the UWB. Severe pattern degradation has also been observed at these frequencies. In RMA, the radiation patterns are found to be stable with minimum degradation at the higher frequencies. The design of RMA has achieved this virtue by manipulating the surface currents in the geometry. Abrupt discontinuities that can cause reflections in the geometry are avoided by the impedance transformer designed in the ground plane.

As frequency domain characterizations hardly throw light into the pulse handling capabilities UWB antennas, a method to characterize them in the time domain is presented in this work. Temporal characterizations of the two antennas performed by the method outlined here reveal a close correspondence between the geometry of the antenna with its performance in the time domain. From the study, it is concluded that for good time-domain performance, excitation of multiple modes within the operating band of the UWB antennas has to be taken care of during the designing of the antennas itself and its physical dimension has to remain to a minimum. The first condition minimizes internal reflections and the subsequent cancelations in the antenna geometry while the second one ensures that there are no changes in the phase center of the antenna.

Acknowledgements

Simulations and measurements of the present work were done at the Govt. Victoria College, Palakkad and Cochin University of Science and Technology, Cochin. The authors acknowledge DST SERB, DST-FIST, UGC and KSCSTE for funding the facilities for simulation and measurement.

The authorship criteria are listed in our Authorship Policy: https://www.intechopen.com/page/authorship-policy.

Conflict of interest

The authors declare no conflict of interest.

Abbreviations

FCC	Federal Communication Commission
FWHM	Full Width Half Maxima
IFFT	Inverse Fast Fourier Transform
RMA	Rectangle Monopole Antenna
SQMA	Square Monopole Antenna
UWB	Ultra Wide-band

Author details

Gopikrishna Madanan[1*†] and Deepti Das Krishna[2†]

1 Department of Physics, Government Victoria College, Palakkad, Kerala, India

2 Department of Electronics, Cochin University of Science and Technology, Cochin, Kerala, India

*Address all correspondence to: gopikrishna@gvc.ac.in

† These authors contributed equally.

References

[1] Ghosh D, De A, Taylor MC, Sarkar TK, Wicks MC, Mokole EL. Transmission and reception by ultra-wideband (UWB) antennas. IEEE Antennas and Propagation Magazine. 2006;48(5):67–99.

[2] Galvan-Tejada GM, Peyrot-Solis MA, Aguilar HJ. Ultra wideband antennas: design, methodologies, and performance. CRC Press; 2017.

[3] Ma TG, Jeng SK. A printed dipole antenna with tapered slot feed for ultrawide-band applications. IEEE Transactions on Antennas and Propagation. 2005;53(11):3833–3836.

[4] Ma TG, Jeng SK. Planar miniature tapered-slot-fed annular slot antennas for ultrawide-band radios. IEEE transactions on Antennas and Propagation. 2005;53(3):1194–1202.

[5] Chu QX, Yang YY. A compact ultrawideband antenna with 3.4/5.5 GHz dual band-notched characteristics. IEEE transactions on antennas and propagation. 2008;56(12):3637–3644.

[6] Acharjee J, Mandal K, Mandal SK, Sarkar PP. A compact printed monopole antenna with enhanced bandwidth and variable dual band notch for UWB applications. Journal of Electromagnetic Waves and applications. 2016;30(15): 1980–1992.

[7] Lee SS, Choi SS, Park JK, Cho KR. Experimental study of UWB antenna in the time domain. Microwave and Optical Technology Letters. 2005;47(6):554–558.

[8] Shameena V, Suma M, Raj Rohith K, Bybi P, Mohanan P. Compact ultra-wideband planar serrated antenna with notch band ON/OFF control. Electronics Letters. 2006;42(23):1323–1324.

[9] Wu Q, Jin R, Geng J, Lao J. Ultra-wideband rectangular disk monopole antenna with notched ground. Electronics Letters. 2007;43(11): 605–606.

[10] Li Y, Li W, Ye Q. A reconfigurable triple-notch-band antenna integrated with defected microstrip structure band-stop filter for ultra-wideband cognitive radio applications. International Journal of Antennas and Propagation. 2013;2013.

[11] Gopikrishna M, Krishna D, Chandran A, Aanandan C. Square monopole antenna for ultra wide band communication applications. Journal of Electromagnetic Waves and Applications. 2007;21(11):1525–1537.

[12] Gopikrishna M, Krishna DD, Anandan C, Mohanan P, Vasudevan K. Design of a compact semi-elliptic monopole slot antenna for UWB systems. IEEE Transactions on Antennas and Propagation. 2009;57(6):1834–1837.

[13] Fereidoony F, Chamaani S, Mirtaheri SA. UWB monopole antenna with stable radiation pattern and low transient distortion. IEEE Antennas and Wireless Propagation Letters. 2011;10: 302–305.

[14] Jongh RK, Hajian M, Ligthart LP. Antenna Time–Domain Measurement Techniques. IEEE Antennas and Propagation Magazine. 1997 October;39 (5):7–12.

[15] Shlivinski A, Heyman E, Kastner R. Antenna characterization in the Time Domain. IEEE Transactions on Antennas and Propagation. 1997;45(7): 1140–1149.

[16] Sorgel W, Knorzer S, Wiesbeck W. Measurement and Evaluation of ultra wideband antennas for communications. In: International ITG Conference on Antennas – INICA2003. Germany; 2003. p. 377–380.

[17] Sorgel W, Weisbeck W. Influence of the antennas on the ultra–wideband transmission. EURASIP Journal on Applied Signal Processing. 2005;3: 296–305.

[18] Zwierzchowski S, Jazayeri P. Derivation and determination of the antenna transfer function for use in ultra–wideband communications analysis. Wireless Proceedings. 2003 July;.

[19] Mohammadian AH, Rajkotia A, Soliman SS. Characterization of UWB transmit–receive antenna system. In: IEEE Conf. Ultra Wideband Systems and Technology; 2003. p. 157–161.

[20] Duroc Y, Ghiotto A, Vuong TP, Tedjini S. UWB Antennas: Systems With Transfer Function and Impulse Response. IEEE Transaction on Antennas and Propagation. 2007 May;55 (5):1449–1451.

[21] Telzhensky N, Leviatan Y. Novel Method of UWB Antenna Optimization for Specified Input Signal Forms by Means of Genetic Algorithm. IEEE Transaction on Antennas and Propagation. 2006 August;54(8): 2216–2225.

Chapter 7

Analysis of Wideband Second-Order Microwave Integrators

Usha Gautam and Tarun Kumar Rawat

Abstract

This chapter presents the implementation of stable, accurate, and wideband second-order microwave integrators (SOMIs). These SOMI designs are obtained by the use of various cascading combinations of transmission line sections and shunt stubs. In order to obtain the optimal values of the characteristic impedances of these line elements, the particle swarm optimization (PSO), cuckoo search algorithm (CSA) and gravitational search algorithm (GSA) are used to approximate the magnitude response of the ideal second-order integrator (SOI). Based on magnitude response, absolute magnitude error, phase response, convergence rate, pole-zero plot, and improvement graph, the performance measure criteria for the proposed SOMIs are performed. The results of the simulation and statistical analysis reveal that GSA exceeds the PSO and CSA in order to approximate the ideal SOI in all state-of-the-art eligible for wide-band microwave integrator. The designed SOMI is compact and suitable for applications covering ultra-wideband (UWB). The designed SOMI structure is also simulated on Advanced Design Software (ADS) in the form of a microstrip line on a dielectric constant 2.2 RT/Duroid substrate with a height of 0.762 mm. In the 3–15 GHz frequency range, the simulated magnitude result agrees well with the ideal one.

Keywords: Cuckoo search algorithm, gravitational search algorithm, particle swarm optimization, line elements, microwave integrator

1. Introduction

Integration plays an important role in many scientific and engineering applications. An integrator is an electronic circuit which produces the output that is the integral of the input applied. Electronic analogue integrators are the basis of analog computers and charge amplifiers, which are performed in the continuous time domain. The integrator is widely used in analog computers, analog-to-digital converters and wave-shaping circuits. Initially, an RC integrator is a circuit that approximates the mathematical process of integration. A simple R-C integrator circuit is shown in **Figure 1**, in which a capacitor (C) in series with a resistor (R) and the source (V_{in}). The output (V_{out}) of the circuit is taken across the capacitor (C).

Let i is the resulting current. Applying Kirchhoff's voltage law to the circuit,

$$V_{in} = iR + \frac{1}{C}\int_0^t i.dt \tag{1}$$

$$V_{out} = \frac{1}{C}\int_0^t i.dt \tag{2}$$

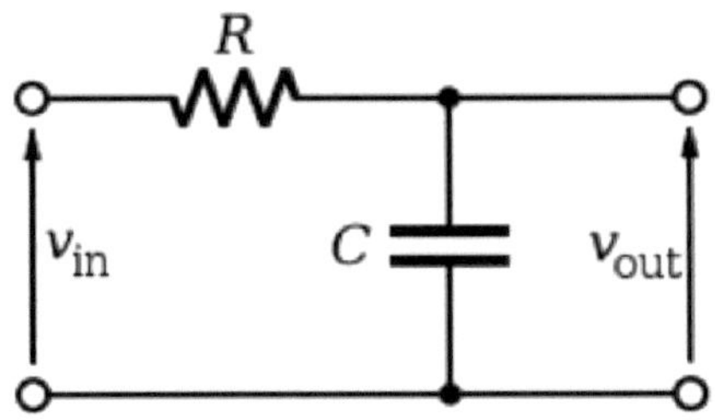

Figure 1.
R-C Integrator circuit.

Multiplying throughout by C, we get

$$CV_{in} = iRC + \int_0^t i.dt \tag{3}$$

as $RC \gg t$, the term $\int_0^t i.dt$ may be neglected

$$CV_{in} = iRC \tag{4}$$

Integrating with respect to t on both sides of Eq. (3)

$$\int_0^t CV_{in} = RC\int_0^t i.dt \tag{5}$$

$$\frac{1}{C}\int_0^t i.dt = \frac{1}{RC}\int_0^t V_{in}.dt \tag{6}$$

From Eq. (2),

$$V_{out} = \frac{1}{RC}\int_0^t V_{in}.dt \tag{7}$$

Eq. (7) shows that the output of an integrator circuit is the integral of the input signal. These analog integrators are limited for low frequency application. Thus, the researcher moved to design digital integrators. Digital integrator is a system that performs mathematical operations on a sampled discrete time signal to reduce or enhance certain aspects of that signal. It is commonly used for applications such as waveform shaping, coherent detection, edge detection, and accumulator analysis in biomedical engineering and signal processing. It is widely utilized in biomedical engineering and signal processing applications, for example, as waveform shaping, coherent detection, edge detection, and accumulator analysis. It is also used in radar applications such as the allocation of mobile satellites, enterprise networks, commercial television services and digital services [1]. In order to design the wideband digital integrators, various methods were intended. Using the Newton-cotes integration rule and various digital integration techniques, the Recursive wideband digital integrators have been designed [2]. For low-speed applications up to barely a few hundred MHz, the integrators are primarily designed and implemented. Therefore, to cover wideband applications such as radar and wireless communication, the design and implementation of integrators for high-frequency applications is necessary. The microwave integrator is essentially used to measure the time integral of the input signal at microwave frequencies (0.3–300 GHz). Using wideband integrators, the high-frequency active filters can be introduced, and these wideband integrators can also be used for industrial and real-time applications for ultra-wideband (frequency range 3.1–10.6 GHz) applications [3–5]. In the z-domain, Hsue et al. have introduced a first-order trapezoidal-rule microwave integrator using a chain-scattering transmission matrix with an operating frequency range from 1 to 10 GHz [6]. By cascading equal-length transmission line sections in the z-domain, three microwave integrators and differentiators have been designed and implemented with different time constants [7, 8]. The second-order microwave integrator (SOMI) was designed by Tsai et al. [9] in the z-domain. A further first-order microwave integrator was designed by Gautam et al. using the ABCD transmission matrix with a bandwidth of 4 to 10 GHz [10]. SOMI was designed by Gupta et al. in the z-domain over the 1.5–5.5 GHz frequency range [11]. Another SOMI was designed by Gautam et al. in the z-domain over the frequency range 3–15 GHz [12]. Nowadays, analyzers are gravitating towards the use of population-based meta-heuristic algorithms to optimize system coefficients in order to design complex or multi - modal systems [13–17]. This chapter introduces modern and compact SOMI designs that lead to wide bandwidth. These designed SOMIs are accomplished by cascading three transmission line sections and two single section stubs of equal length. By population-based meta heuristic algorithms, the optimum value of characteristic impedances of these line elements are obtained. A global cost-function solution is achieved by minimizing the error gap between the ideal second-order integrator (SOI) and the designed SOMIs. The design-1 SOMI approximates the ideal SOI over the 2.5 to 16 GHz frequency range, and the design-2 SOMI approximates the ideal SOI over the 3 to 15 GHz frequency range. These designed wideband SOMIs would operate with a wider frequency band to be used in mobile communication on a mobile network such as 4 G and 5 G (above 3 GHz) [18]. All the simulated results are obtained by MATLAB and ADS. These simulated outcomes are formulated to be in close agreement with the ideal one. The novelty of these designed SOMIs exists in terms of wide bandwidth and miniaturization of hardware.

2. Problem formulation of SOMIs

To realize second-order microwave integrator in microwave range, consider a two-port network, which is illustrated in **Figure 2**.

Its scattering matrix is defined as [4].

$$\begin{bmatrix} b_1 \\ b_2 \end{bmatrix} = \begin{bmatrix} S_{11} & S_{12} \\ S_{21} & S_{22} \end{bmatrix} \begin{bmatrix} a_1 \\ a_2 \end{bmatrix} \tag{8}$$

where a_1 and a_2 are incident waves at port 1 and port 2, respectively, and b_1 and b_2 are reflected waves at port 1 and port 2, respectively [4]. The chain scattering matrix of a two-port network can be established from the scattering matrix (S-matrix). The chain-scattering matrix of two-port network is defined as [5].

$$\begin{bmatrix} a_1 \\ b_1 \end{bmatrix} = \begin{bmatrix} T_{11} & T_{12} \\ T_{21} & \mathrm{T}_{22} \end{bmatrix} \begin{bmatrix} b_2 \\ a_2 \end{bmatrix} \tag{9}$$

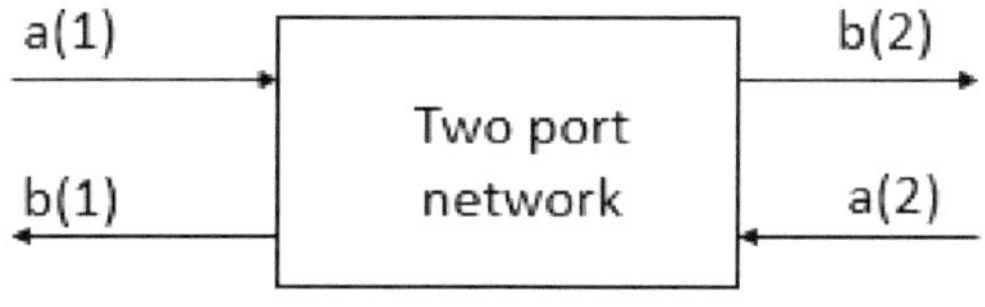

Figure 2.
Two-port network.

For network computation the chain-scattering matrix from S-matrix (on solving Eq. (8) and Eq. (9))

$$T_{11} = \frac{1}{S_{21}}$$

$$T_{12} = -\frac{S_{22}}{S_{21}}$$

$$T_{21} = \frac{S_{11}}{S_{21}}$$

$$T_{22} = \frac{S_{12}\,S_{21} - S_{11}S_{22}}{S_{21}}$$

Then the chain-scattering matrix is

$$\begin{bmatrix} T_{11} & T_{12} \\ T_{21} & T_{22} \end{bmatrix} = \begin{bmatrix} \frac{1}{S_{21}} & -\frac{S_{22}}{S_{21}} \\ \frac{S_{11}}{S_{21}} & \frac{S_{12}\,S_{21} - S_{11}S_{22}}{S_{21}} \end{bmatrix} \tag{10}$$

The formulation of SOMIs is employed with equal length line elements is cascading. The overall transfer function of a cascaded network can be established by multiplying the chain scattering matrices of the line elements. These line elements can be transmission line sections and stubs. Assume the length of all transmission line sections and stubs is $l = \lambda_o/4$, where λ_o represents the wavelength of the lines at the normalizing angular frequency or we can say that the electrical length of each section (stubs and transmission lines) is set to be 90° at the normalizing frequency. The electrical length of line element is $\theta = \beta l$.

The frequency response of an ideal second order integrator (SOI) is given by

$$H(j\omega) = \frac{1}{\omega^2} \tag{11}$$

where ω represents the angular frequency in radians per second. Since the magnitude response of an open-circuited stub decreases with frequency, an open-circuited stub can be chosen to design an integrator. Assume the impedance of an open-circuited stub to be Z_{oc}. Its chain scattering is [5].

$$\begin{bmatrix} T_{11} & T_{12} \\ T_{21} & T_{22} \end{bmatrix}_{open} = \begin{bmatrix} 1+j\frac{Z_0}{2Z_{oc}}\tan(\beta l) & j\frac{Z_0}{2Z_{oc}}\tan(\beta l) \\ -j\frac{Z_0}{2Z_{oc}}\tan(\beta l) & 1-j\frac{Z_0}{2Z_{oc}}\tan(\beta l) \end{bmatrix} \tag{12}$$

where Z_0 represents the reference characteristic impedance that is 50 Ω, Z_{oc} is the characteristic impedance of an open circuited stub and β represents the phase constant.

Assume ω is the angular frequency and τ is the propagation delay attributable to the length l. Therefore, the term $j\tan(\beta l) = j\tan(\omega\tau)$ can be expressed by $D^{-1} = e^{-j\omega\tau}$, which can be considered as a unit of delay, that is [5].

$$\text{jtan}(\omega\tau) = \frac{e^{j\omega\tau} - e^{-j\omega\tau}}{e^{j\omega\tau} + e^{-j\omega\tau}} = \frac{D - D^{-1}}{D + D^{-1}} \tag{13}$$

Subsequently, the chain scattering matrix of open-circuited stub is

$$\begin{bmatrix} T_{11} & T_{12} \\ T_{21} & T_{22} \end{bmatrix}_{open} = \frac{1}{1+D^{-2}} \begin{bmatrix} (1+k)+(1-k)D^{-2} & (k-kD^{-2}) \\ -k+kD^{-2} & (1-k)+(1+k)D^{-2} \end{bmatrix} \tag{14}$$

where $k = \frac{Z_0}{2Z_{oc}}$. Now we substitute $z = D^2$ in above Eq. (14) for designing purpose, which resembles a scaling by two on frequency axis. Then the chain scattering matrix of open-circuited stub is [6].

$$\begin{bmatrix} T_{11} & T_{12} \\ T_{21} & T_{22} \end{bmatrix}_{open} = \frac{1}{1+z^{-1}} \begin{bmatrix} (1+k)+(1-k)z^{-1} & (k-kz^{-1}) \\ -k+kz^{-1} & (1-k)+(1+k)z^{-1} \end{bmatrix} \tag{15}$$

and S_{21} is given by

$$S_{21} = \frac{1}{T_{11}} = \frac{1+z^{-1}}{(1+k)+(1-k)z^{-1}} \tag{16}$$

where $z = e^{j\beta l}$. Likewise, the chain scattering matrix for the transmission line section in Z domain is given by [6].

$$\begin{bmatrix} T_{11} & T_{12} \\ T_{21} & T_{22} \end{bmatrix}_{TLS} = \frac{1}{z^{-\frac{1}{2}}(1-\Gamma^2)} \begin{bmatrix} 1-\Gamma^2 z^{-1} & -\Gamma-\Gamma z^{-1} \\ 1-\Gamma z^{-1} & -\Gamma^2+z^{-1} \end{bmatrix} \tag{17}$$

where the reflection coefficient Γ is given by

$$\Gamma = \frac{Z_{TL}-Z_0}{Z_{TL}+Z_0} \tag{18}$$

where Z_{TL} represents the characteristic impedance of serial transmission line section. Similarly, the chain scattering matrix of a short-circuited stub is given by [8]

$$\begin{bmatrix} T_{11} & T_{12} \\ T_{21} & T_{22} \end{bmatrix}_{Short} = \frac{1}{1-z^{-1}} \begin{bmatrix} (1+\delta)-(1-\delta)z^{-1} & (\delta+\delta z^{-1}) \\ -\delta-\delta z^{-1} & (1-\delta)-(1+\delta)z^{-1} \end{bmatrix} \tag{19}$$

where the coefficient δ is given by

$$\delta = \frac{Z_0}{2Z_{SC}} \tag{20}$$

where Z_{SC} represents the characteristic impedance of the short-circuited stub. Serial line sections and short-circuited stubs shunted with open circuited stub should be employed in the transmission line configuration to design an integrator by cascading. A cascaded connection of two-port network is equivalent to a single two-port network containing a product of matrices. Assume the SOI is composed of P open-circuited stubs, Q short-circuited stubs and R transmission line sections. The overall S_{21} in generalized form is given by

$$S_{21} = \frac{1}{T_{11}} = \frac{(1+z^{-1})^P (1-z^{-1})^Q z^{-R/2} \prod_{r=1}^{R} (1-\Gamma_r^2)}{\alpha_o + \alpha_1 z^{-1} + \alpha_2 z^{-2} + \alpha_3 z^{-3} + \ldots + \alpha_N z^{-N}} \tag{21}$$

where, P denotes the number of open-circuited stubs, Q denotes the number of short-circuited stubs, R denotes the number of transmission line section, N is the total number of line elements $N = P + Q + R$ and α_o, α_1, α_2, ..., α_N Coefficients are functions of the reflection coefficient of each section of the transmission line and stub characteristic impedances (in terms of open-circuited stub coefficient k and short-circuited stub coefficient δ). In addition, to design the SOMI, serial transmission lines cascaded with shunt circuited stubs can be employed. Then the frequency-domain response of the transfer function, which has an integrator characteristic, is thus obtained. In order to design a wideband SOMI, the following two methods are used.

Design-I (SOMI using two open stubs and three transmission line sections): In this design, we may select two open-circuited stubs and three transmission line sections as shown in **Figure 3**, where $P = 2, R = 3$. Overall S_{21} of the design-1 SOMI is given by

$$S_{21} = \frac{\left(1 + z^{-1}\right)^2 z^{-3/2} \prod_{r=1}^{3} \left(1 - \Gamma_r^2\right)}{\alpha_o + \alpha_1 z^{-1} + \alpha_2 z^{-2} + \alpha_3 z^{-3} + \alpha_4 z^{-4} + \alpha_5 z^{-5}} \tag{22}$$

Design-II (SOMI using one open stub, one short stub and three transmission line section): In this design, a short-circuited stub is cascaded with the transmission line sections and an open-circuited stub as shown in **Figure 5**, where $P = 1, Q = 1$, $R = 3$. Overall S_{21} of the design-2 SOMI is given by

$$S_{21} = \frac{\left(1 - z^{-2}\right) z^{-3/2} \prod_{r=1}^{3} \left(1 - \Gamma_r^2\right)}{\alpha_o + \alpha_1 z^{-1} + \alpha_2 z^{-2} + \alpha_3 z^{-3} + \alpha_4 z^{-4} + \alpha_5 z^{-5}} \tag{23}$$

The term $(1 + z^{-1})$ is due to open-circuited stub, the term $(1 - z^{-1})$ is due to short-circuited stub, and $z^{-3/2}$ is the delay factor of the transmission line section. These line elements are transmission lines of equal length with a length of $l = \lambda_0/4$ at an operating frequency of 12.5 GHz.

The next task be to achieve the optimum value of characteristic impedances of design-1 and design-2 SOMI line elements. The optimization algorithms are used to obtain these characteristic impedances of the line elements. In order to lower the cost function, the design of SOMI is considered as an approximation problem. The cost function differs between the response in magnitude of an ideal SOI and the SOMI designed. The cost function is formulated in the sense of least squares and can be expressed as

$$CF = min \int_0^{\pi} |E(\omega)|^2 \, d\omega = min \int_0^{\pi} |H(\omega) - S_{21}|^2 \, d\omega \tag{24}$$

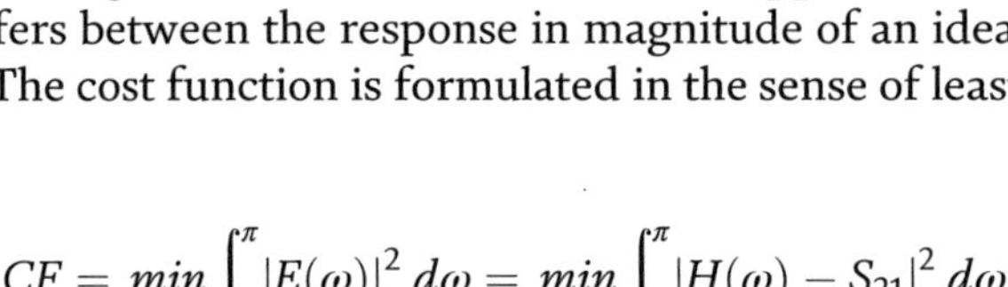

Figure 3.
Configuration of the design-1 SOMI.

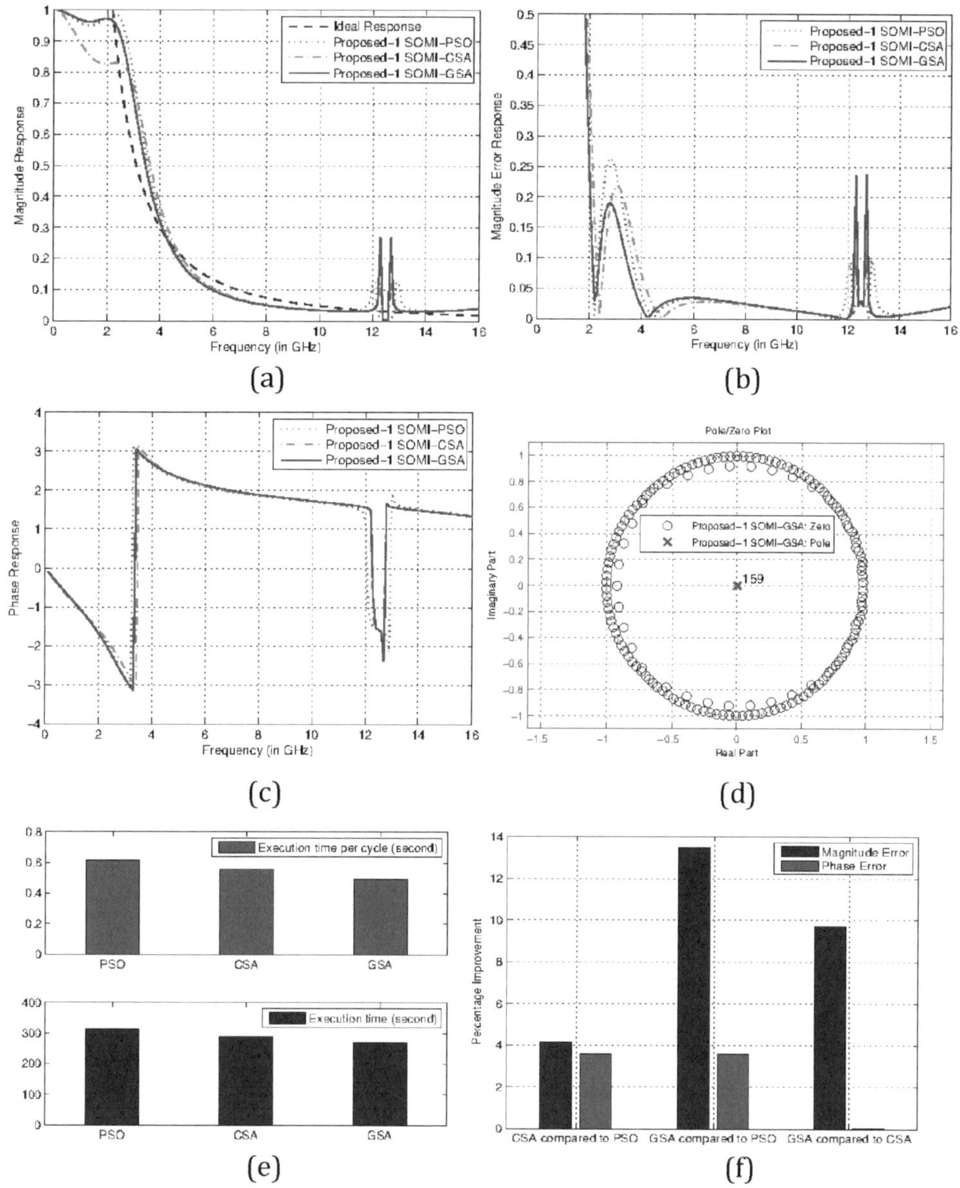

Figure 4.
Profile of design-1 SOMI (a) Magnitude response, (b) Magnitude error response, (c) Phase response, (d) Pole-zero plot, (e) Convergence profile, and (f) Improvement bae-graph.

where $E(\omega) = (H(\omega) - S_{21})$ is the cost function, in which $H(\omega)$ is the frequency response of an ideal SOI.

3. Employed optimization methods

In infinite impulse response (IIR) systems, the error surface is generally non-quadratic and multimodal with respect to the system parameters. Minimization of such error fitness function using derivative-based search algorithm is difficult. This is due to the fact that the derivative-based search algorithm may not converge to the global minima and get stuck in local minima. Moreover, IIR systems are associated with the stability issues as the poles of the systems may lie outside the unit circle. Such techniques are found unfit to solve multi-objective, multi-modal complex problems and a fine tuning of algorithm parameters is required. To conquer these disadvantages, several practitioners rely on meta-heuristic algorithms, which are based on natural evolution. The meta-heuristic algorithms are nature inspired population-based search techniques which have the ability to serve a global optimal solution with high convergence by accumulating random search and selection principle. Therefore, intelligent search paradigms and optimization methods are adopted in this work for an optimal differentiator design (a multi-modal problem) in short computation time and with high accuracy. Three population-based heuristic search algorithms PSO, CSA, and GSA are employed in this section to diminish the cost function in order to find optimum values of characteristic impedances for designed wideband SOMIs of line elements. **Table 1** displays the optimum set of algorithm control parameters for designed SOMIs. A brief description of all three algorithm is discussed below.

3.1 Particle swarm optimization

Particle swarm optimization was proposed by Kennedy and Eberhart in nineties based on swarm behavior. PSO has simplicity, high solution quality and superior convergence characteristics as compared to other algorithms. It is more efficient, easy to implement and flexible to control between global and local exploration of the search space [13, 14]. In PSO, every particle has a candidate solution and each candidate solution has its position and velocity. The position vector gives the required solution and the velocity vector gives the current position by which it reaches at the new position. After every iteration, the position and the velocity vectors are updated unless final coefficients are obtained. The velocity and position vector are updated according to the following equations:

Parameters	PSO	CSA	GSA
Population size	n_p =25	n_c =25	n_g =25
Maximum Iteration	400	400	400
C_1	2	—	—
C_2	2	—	—
v_i^{min}	0.01	—	—
v_i^{max}	1.0	—	—
w_{min}	0.1	—	—
w_{max}	1.0	—	—
Discovering rate of alien egg, Pa	—	0.25	—
G_0	—	—	100
α	—	—	20
ϵ	—	—	0.001
r^{Norm}	—	—	2
r^{Power}	—	—	1

Table 1.
Control parameters of PSO, CSA and GSA for optimization.

$$v_i^{(k+1)} = w * v_i^{(k)} + C_1 * rand1 * \left(gbest_i^{(k)} - p_i^{(k)}\right) + C_2 * rand2 * \left(pbest_i^{(k)} - p_i^{(k)}\right) \tag{25}$$

$$p_i^{(k+1)} = p_i^{(k)} + v_i^{(k+1)} \tag{26}$$

where, v_i^k is the velocity of the i *th* particle and ω is the weight factor. C_1 and C_2 are positive cognitive parameters while $rand1$ and $rand2$ are two random parameters has range from $[0, 1]$. $gbest^k$is the global best position component at k *th* iteration and $pbest_i^{(k)}$ is the particular best value of i *th* particle. $pbest_i^{(k)}$ is the position vector of i *th* particle at k *th* iteration. Some parameters are selected to obtain the characteristic impedances of line elements, which are provided in **Table 1**.

3.2 Cuckoo search algorithm

The cuckoo search algorithm is developed by Yang and Deb in 2009 which is inspired by the concept of unique breeding behavior of cuckoo bird in combination with Lèvy (λ) fights. The theoretical concepts of CSA are well developed and tested in [15–17]. The single parameter setting in CSA is proven to be a crucial superiority factor as compared to other nature-based algorithms. As a result, the optimized results are executed in very less time. CSA is employed in this section for determining the optimum characteristic values of impedances of transmission line elements on account of minimizing the magnitude error of designed SOMI.

The species of cuckoo birds lay eggs in other bird nests, where the host birds either throw off the detected strange eggs or leave their nests and move into a new spot. The algorithm symbolizes each host nest to a capable solution for this design problem and assigns it a fitness value, as defined in Eq. (24). Furthermore, CSA starts to exchange the current fitness value with a better solution iteratively. The concept of Lèvy flights is then introduced in the process for exploration of new solutions, mathematically modelled using the Lèvy distribution, $1 < \lambda \leq 3$ [17] with an infinite variance and infinite mean.

3.3 Gravitational search algorithm

This algorithm is based on Newton's theory of gravitational force and was introduced by E Rashedi et al. in 2009 [19]. Candidates in GSA are supposed to be objects with a given mass. In accordance with Newton's Law, every candidate in a space has an attraction force with every other candidate. This force correlates inversely to the square of the distance between the candidates and is directly proportional to the product of their masses [20, 21]. Each candidate in the GSA has certain parameters: position of candidate's mass, inertial mass of candidate, active gravitational mass and passive gravitational mass between candidates. The position of the mass of candidates resembles the solution of the problem devised [21]. The gravitational mass and inertial mass are determined using a cost function of designed problem. The mathematical representation of GSA is considering $\boldsymbol{N}$ candidates within a system and all candidates are randomly positioned in search space. The gravitational force on candidate $\boldsymbol{i}$ from candidate $\boldsymbol{j}$ at time $\boldsymbol{t}$ in dimension $\boldsymbol{d}$ is defined as [19].

$$F_{ij}^{d(t)} = G^{(t)} \frac{M_{pi}^{(t)} \times M_{aj}^{(t)}}{R_{ij}^{(t)} + \varepsilon} \left(x_j^{d(t)} - x_i^{d(t)}\right) \tag{27}$$

where, $G^{(t)}$ represents gravitational constant at time t, $M_{pi}^{(t)}$ represents the passive gravitational mass related to candidate i, $M_{aj}^{(t)}$ represents the active gravitational mass related to candidate j, $R_{ij}^{(t)}$ represents the Euclidian distance between two candidates i and j and ε is a small constant. Then $G^{(t)}$ is calculated as [19].

$$G^{(t)} = G_0 \times \exp\left(-\alpha \times itr/maxitr\right) \tag{28}$$

where, G_0 represents the initial value of gravitation constant, α is descending coefficient, the current iteration is represented by itr, and $maxitr$ represents the maximum number of iterations.

Then the total force that acts on candidate i in dimension d is calculated as

$$F_i^{d(t)} = \sum_{j=1,\ j\neq 1}^{N} rand_j F_{ij}^{d(t)} \tag{29}$$

where $rand_j$ is a random number. As reported to the motion's law, the acceleration of candidate i is given as

$$acc_i^{d(t)} = \frac{F_i^{d(t)}}{Ma_{ii}^{(t)}} \tag{30}$$

where Ma_{ii} represents the mass of the candidate i. The inertial mass and the gravitational mass are updated as follows [19].

$$m_i^{(t)} = \frac{CF_i^{(t)} - worst^{(t)}}{best^{(t)} - worst^{(t)}} \tag{31}$$

where CF represents the cost function of the candidate i.

$$M_{pi}^{(t)} = \frac{m_i^{(t)}}{\sum_{j=1}^{J} m_j^{(t)}} \tag{32}$$

The best and worst value are given by

$$best^{(t)} = min_{\ j\in(I..J)} CF_j^{(t)} \tag{33}$$

$$worst^{(t)} = max_{\ j\in(I..J)} CF_j^{(t)} \tag{34}$$

Then the position and the velocity of candidates are given by [19].

$$position : x_i^{d(t+1)} = x_i^{d(t)} + v_i^{d(t+1)} \tag{35}$$

$$velocity : v_i^{\mathrm{d}(t+1)} = rand_i \times v_i^{d(t)} + acc_i^{d(t)} \tag{36}$$

where $rand_i$ represents a random number. Finally, the position and velocity of the candidate are obtained.

Algorithm 1
Pseudo code for GSA for the design of SOMI-GSA
Define $H(\omega)$ and fitness function, Eq. (24)
Initialize population size of candidates, ng and other control parameters
Set upper and lower bounds, maximum iterations, generate population
While iteration, l increases, $N < 400$ do
Compute fitness using Eq. (24)
Evaluate gravitational constant, G_0 and *gbest*
Compute masses, gravitational forces, and acceleration using Eqs. (27)–(32). for each candidate
Updae the velocity and position of each candidate using Eqs. (35), (36)
For minimzation,
If $E^{l+1} < E^{l}$
New solutions are updated
EndIf
EndWhile
Record the best solution
Characteristic impedances optimized

4. Simulation results

In this section, simulation results are discussed and analyzed. All the simulation results are carried out in MATLAB environment. The same control parameters for PSO, CSA and GSA have been selected for appropriate comparison of optimization algorithms. The lower and upper limits of the optimized coefficients are set to be 10 and 150 for functional realizability. Absolute magnitude error (AME), phase response, pole-zero plot, convergence rate and improvement rate are taken into account in assessing the performance of the proposed SOMI magnitude response.

4.1 Design-1 [SOMI using two open stub and three serial line sections]

Figure 3 presents the configuration of the design-1 SOMI, and **Table 2** displays the characteristic impedances of design-1 SOMI achieved with PSO, CSA and GSA. The algorithm steps have shown in section-3 by which these characteristic impedances are attained. The magnitude response, the AME, the phase response, the pole-zero plot, the fitness rate and the improvement graph are the parameters selected in **Figure 4(a)–(f)** for the frequency response analysis of the design-1 SOMI. **Table 3**

Characteristic impedance	PSO	CSA	GSA
Z_{oc1}	12	112.2	148.8
Z_{TL1}	49	17	10
Z_{TL2}	120	114	149
Z_{TL3}	14	11	43
Z_{oc2}	115	14	21

Table 2.
Optimized characteristic impedances of the design-1 SOMI.

Algorithm	Total magnitude error	Phase error
PSO	5.7746	1.0403
CSA	5.5353	1.0031
GSA	4.9954	1.0030

Table 3.
Comparison summary of total magnitude and phase error for design-1 SOMI.

records the overall magnitude error and the phase error. **Table 4** and **Table 5** summarize the statistical analysis and the qualitatively analyzed data of magnitude error of the design-1 SOMI. It is observed that GSA-based results have least values in all aspects with respect to PSO and CSA. The pole-zero plot of the Design-1 SOMI for stability analysis in which all the poles and zeros lie inside the unit circle is shown in **Figure 4(d)**. This plot asserts that the SOMI design-1 is stable. The convergence profile of all three algorithms in which GSA is clearly demonstrated to be faster than PSO and CSA in less execution time is shown in **Figure 4(e)**, which is also summarized in **Table 6**. The percentage improvement comparison in magnitude error for the design-1 SOMI as CSA to PSO, GSA to PSO, and GSA to CSA and the percentage improvement comparison in phase errors summarized in **Table 7** are shown in **Figure 4(f)**. The design-1 SOMI GSA-based magnitude response closely matches that of the ideal one in the 2.5–16 GHz frequency range. On the basis of these observations, compared to the PSO and CSA, GSA has the lowest magnitude error and the highest convergence speed.

Algorithm	Mean	Variance	Standard deviation
PSO	0.0410	0.0035	0.0588
CSA	0.0393	0.0036	0.0603
GSA	0.0354	0.0024	0.0494

Table 4.
Statistical data of magnitude error for design-1 SOMI.

Algorithm	Maximum	Minimum	Average
PSO	0.2642	4.8442×10^{-4}	0.0410
CSA	0.3762	9.2306×10^{-4}	0.0393
GSA	0.2384	4.2118×10^{-4}	0.0354

Table 5.
Qualitatively data of magnitude error for design-1 SOMI.

Algorithm	Iteration cycle	Execution time (s) per iteration	Execution time (s) per cycle
PSO	400	313.2654	0.6132
CSA	400	291.3719	0.5547
GSA	400	269.6370	0.4950

Table 6.
Convergence profile for design-1 SOMI.

Algorithm	Magnitude PI (%)	Phase PI (%)
PSO	4.1613	3.5758
CSA	13.4935	3.5855
GSA	9.7374	0.0099

Table 7.
Percentage improvement in magnitude and phase error for design-1 SOMI.

4.2 Design-2 [SOMI using one open stub, one short stub and three serial lines]

The design-2 SOMI configuration is illustrated in **Figure 5**. The optimized characteristic impedances of design-2 SOMI obtained by PSO, CSA, and GSA. The algorithm steps have discussed in Section 3 by which these characteristic impedances are attained. **Table 8** offers the optimized characteristic impedances of the line elements of Design-2 SOMI. For the frequency response analysis of the design-2 SOMI, the same parameters are selected as the design-1 SOMI as shown in **Figure 6 (a)–(f)**. In **Table 9**, the total magnitude and phase errors are listed. **Table 10** and **Table 11** summarize the statistical analysis and the qualitatively analyzed data of different characteristics of magnitude error of the design-2 SOMI. It is observed that GSA-based results have the least values in all aspects as compared to the PSO and CSA. The pole-zero plot of the Design-2 SOMI for stability analysis in which all the poles and zeros lie inside the unit circle is shown in **Figure 6(d)**. This plot asserts that SOMI Design-2 is stable. The convergence profile of all three algorithms in which GSA is clearly demonstrated to be faster than PSO and CSA with less execution time is shown in **Figure 6(e)**, which is also summarized in **Table 12**. The percentage improvement comparison in magnitude error for the design-2 SOMI as CSA to PSO, GSA to PSO, and GSA to CSA and the percentage improvement comparison in process error, summarized in **Table 13**, are shown in **Figure 6(f)**.

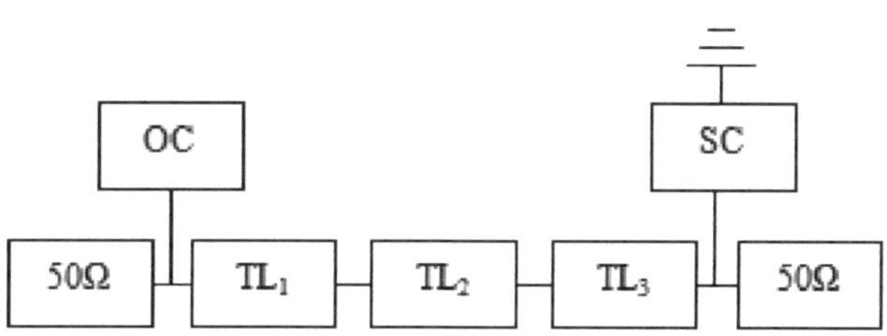

Figure 5.
Configuration of the design-2 SOMI.

Characteristic impedance	PSO	CSA	GSA
Z_{OC}	116	26.87	141
Z_{TL1}	30.5	142.6	33
Z_{TL2}	145	84	138
Z_{TL3}	9.5	19	10
Z_{SC}	110	138.09	116.89

Table 8.
Optimized characteristic impedances of the design-2 SOMI.

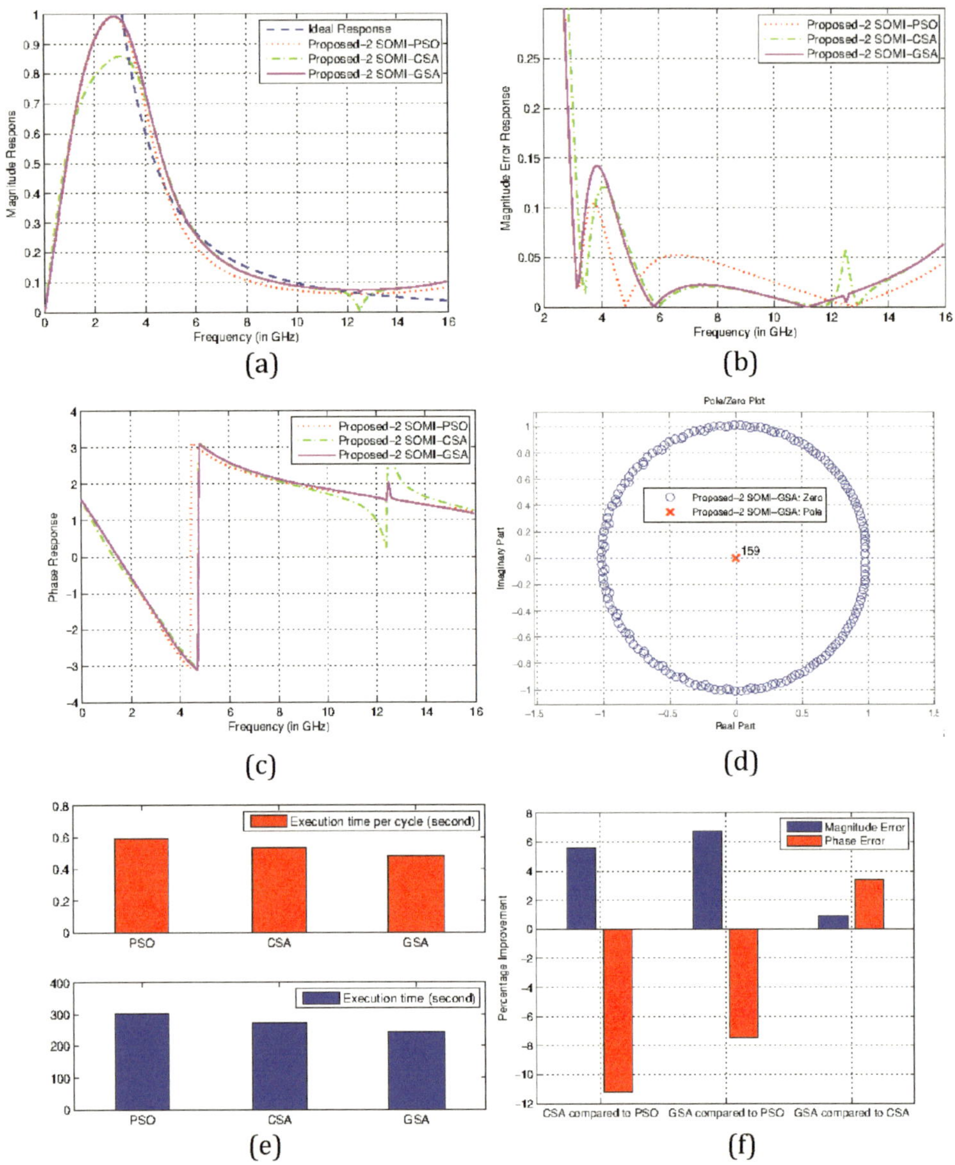

Figure 6.
Profile of the design-2 SOMI (a) Magnitude response, (b) Magnitude error response, (c) Phase response, (d) Pole-zero plot, (e) Convergence profile, and (f) Improvement bar graph.

Algorithm	Total magnitude error	Phase error
PSO	3.9217	0.9861
CSA	3.6920	1.0968
GSA	3.6573	1.0595

Table 9.
Comparison summary of total magnitude and phase error for design-2 SOMI.

The Design-2 SOMI GSA-based magnitude response closely matches that of the ideal one in the 3.0–15 GHz frequency range. Based on the observations, compared to PSO and CSA, GSA has the least magnitude error and highest convergence speed.

Algorithm	Mean	Variance	Standard deviation
PSO	0.0324	5.6118×10^{-4}	0.0237
CSA	0.0303	0.0013	0.0354
GSA	0.0300	0.0012	0.0343

Table 10.
Statistical data of magnitude error for design-2 SOMI.

Algorithm	Maximum	Minimum	Average
PSO	0.1048	2.59×10^{-4}	0.0324
CSA	0.2097	9.35×10^{-4}	0.0303
GSA	0.1425	3.67×10^{-4}	0.0300

Table 11.
Qualitatively data of magnitude error for design-2 SOMI.

Algorithm	Iteration cycle	Execution time (s) per iteration	Execution time (s) per cycle
PSO	400	301.5411	0.5918
CSA	400	273.1862	0.5356
GSA	400	244.6284	0.4802

Table 12.
Convergence profile for design-2 SOMI.

Algorithm	Magnitude PI (%)	Phase PI (%)
PSO	5.8571	−11.22
CSA	6.7419	−7.443
GSA	0.9398	3.400

Table 13.
Percentage improvement in magnitude and phase error for design-2 SOMI.

Figure 7 depicts the comparison of the designed SOMIs with the existing second order microwave integrator. **Figure 7(a)** and **(b)**, respectively, display the comparative magnitude response and AME of the designed SOMI GSA-based optimization with an existing integrator. The corresponding qualitative and statistical magnitude error analysis of the existing SOMIs and the designed integrator is shown in **Table 14**. **Figure 7** and **Table 14** confirm that GSA-based designed SOMIs are more suitable in terms of magnitude, minimal magnitude error, and wide bandwidth, especially for high-frequency ranges.

The results of the proposed-2 SOMI GSA-based leads in performance to the proposed-1 SOMI counterpart. All simulation results of the design-2 SOMI approximates with the ideal one and have linear phase response in its frequency range 3.0–15 GHz. Therefore, the designed-2 SOMI is simulated on ADS by using microstrip lines. To simulate the designed-2 SOMI on ADS, RT/duroid substrate is selected with dielectric constant ϵ_r=2.2, height of substrate $(H) = 30\ mil\ (0.762\ mm)$, loss tangent $\tan(\delta) = 0.001$, and copper cladding is 35 um. The magnitude and phase response of the designed-2 SOMI GSA-based is illustrated in **Figure 8(a)** and **(b)** respectively.

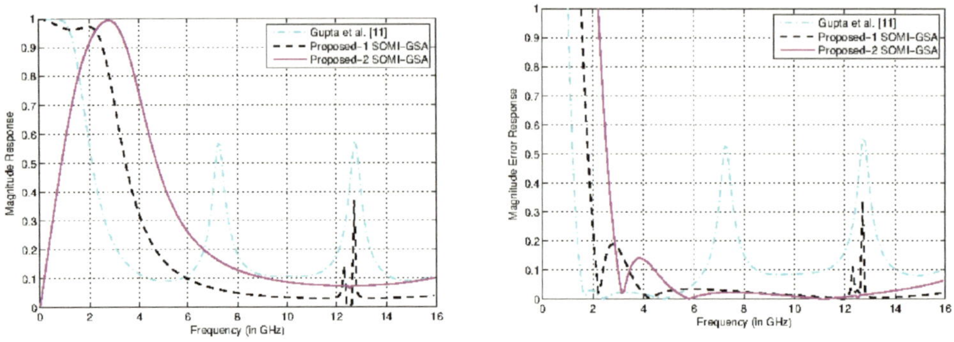

(a) (b)

Figure 7.
Comparison of the designed SOMI-GSA with existing integrator (a) Magnitude reponse and (b) Error response.

Integrator Order	Frequency range	Reference	Total error	Max error	Min error	Mean error	Variance	Standard deviation
2nd	1–10 GHz	Gupta et al. [11]	10.46	1.11	1.756×10^{-4}	0.115	0.0308	0.1755
	1–15 GHz	[11]	17.98	1.11	1.756×10^{-4}	0.137	0.0278	0.1667
2nd	2.5–16 GHz	Design-1 SOMI	4.995	0.238	4.211×10^{-4}	0.035	0.0024	0.0494
2nd	3–15 GHz	Design-2 SOMI	3.657	0.142	3.67×10^{-4}	0.030	0.0012	0.0343

Table 14.
Comparison of magnitude error of the designed SOMIs with existing design method.

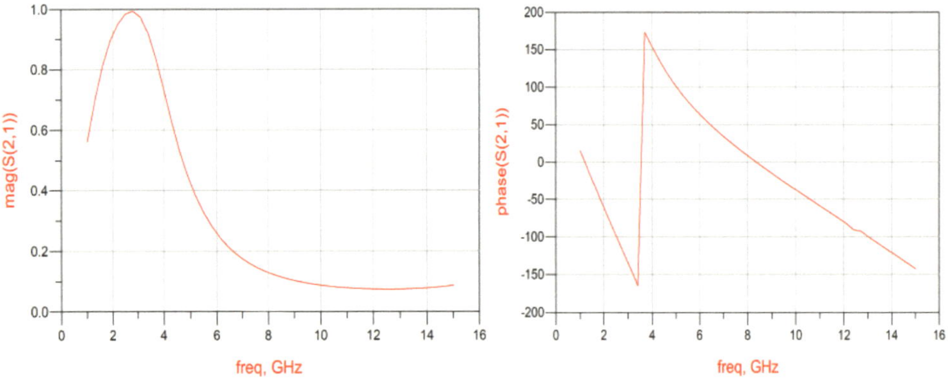

(a) (b)

Figure 8.
Using ADS software (a) Magnitude response and (b) Phase response.

5. Conclusion

The study focused on the design and analysis of compact, stable and wideband second order microwave integrators. The designs are obtained by cascading the line elements i.e. transmission line sections, open-circuited stub and short-circuited stub. The optimum values of line elements are obtained by applying the PSO, CSA and GSA by which the magnitude response of designed integrators approximate the ideal magnitude response. The results are simulated statistically on MATLAB, which affirms that GSA outperforms the PSO and CSA in all state-of-the-art in terms of magnitude response. Furthermore, the designed-2 SOMI GSA-based is also simulated on ADS using microstrip lines. With the exception of lower frequency range (under 3 GHz), the simulated magnitude response of integrator has consistency over the frequency range from 3 to 15 GHz with ideal one in both MATLAB and ADS environment, which makes it appropriate for ultra-wideband applications.

Author details

Usha Gautam* and Tarun Kumar Rawat
Division of Electronics and Communication, Netaji Subhash Institute of Technology, Delhi, India

*Address all correspondence to: usha.aitpg@gmail.com

References

[1] Oppenheim AV and Shafer RW: Discrete-Time Signal Processing. New Jersey: Prentice-Hall. 1989.

[2] Ngo NQ: A new approach for the design of wideband digital integrator and differentiator. IEEE Transactions on Circuits and Systems II: Express Briefs; 2006; 53; 936–940.

[3] Skolink MI: Introduction to Radar Systems. New York: McGrawHill. 1980.

[4] Pozar DM: Microwave Engineering, 3rd ed., Wiley, Singapore, 2005.

[5] Chiang-Hsue: Design and Implementation of filters using transfer function in the Z Domain. IEEE Trans on Microwave theory and Techniques. May 2001; 49; 5.

[6] Hsue CW, Tsai LC and Kan ST: Implementation of a trapezoidal rule microwave integrator. Microwave and Optical Technology Letters. 2006; 48; 822–825.

[7] Hsue C-W, Tsai L-C, Tsai Y-H.: Time-constant control of microwave integrators using transmission lines. IEEE Transactions on Microwave Theory and Techniques. 2006; 54; 1043–1047.

[8] Tsai L-C and Wu Y-T: Time-constant control analysis of microwave differentiators. IET Microwaves, Antennas & Propagation. 2008; 3; 1044–1050.

[9] Tsai LC and Fang HS: Design and implementation of second-order microwave integrators. Microwave and Optical Technology Letters. 2011; 53; 1983–1986.

[10] Gautam U, Upadhyay DK and Rawat TK: New designs of first order microwave integrator. IEEE International Conference on Signal Processing (SPIN), India; 2016 285–289.

[11] Gupta M and Upadhyay DK: New design of second order microwave integrator. IEEE International Conference on Innovative Mechanisms for Industry Applications (ICIMIA), India; 2017; 556–560.

[12] Gautam U, Rawat, TK: The optimal design and analysis of wideband second order microwave integrator. International Journal of Microwave and Wireless Technologies. 2019; 11; 3; 227–236.

[13] Kennedy J and Eberhart R: Particle Swarm optimization. Proceedings of IEEE International Conference Neural Network; 1995; 4;1942–1948.

[14] Shi JY and Eberhart RC: Empirical study of particle swarm optimization. Proceedings of the Congress on Evolutionary Computation (CEC99), Washington, DC, USA; 1999; 3; 1945–1950.

[15] Yang XS and Deb S: Cuckoo search via Lvy flights. Proceedings of world congress on nature and biologically inspired computing. USA IEEE Publications. 2009; 210–214.

[16] Yang XS and Deb S: Cuckoo search: recent advances and applications. Neural Computing & Applications. 2014; 24; 169–174.

[17] Kumar M and Rawat TK: Optimal fractional delay-IIR filter design using cuckoo search algorithm. ISA Transaction. 2015; 59; 39–54.

[18] Hong JS and Lancanter MJ: Microstrip Filters for RF/Microwave Applications. New York: Wiley. 2001.

[19] Rashedi E, Nezamabadi S and Saryazdi S: GSA: a gravitational search algorithm. Information Sciences. 2009; 179; 2232–2248.

[20] Newton Isaac. In Experimental Philosophy Particular Propositions are Inferred from The Phenomena and Afterwards Rendered General by Induction, 3rd ed.: Andrew Motte's English Translation published. 1729.

[21] Rashedi E, Nezamabadi S and Saryazdi S: Filter modelling using gravitational search algorithm. Engineering Applications of Artificial Intelligence's. 2011; 24; 117–122.